THE JERUSALEM-HARVARD LECTURES

*Sponsored by the Hebrew University of Jerusalem
and Harvard University Press*

HOW TO WIN THE NOBEL PRIZE

An Unexpected Life in Science

❊

J. MICHAEL BISHOP

Harvard University Press
Cambridge, Massachusetts
London, England
2003

Frontispiece: William Blake, *Newton,* 1795. © Tate, London 2002.

Library of Congress Cataloging-in-Publication Data

Bishop, J. Michael, 1936–
How to win the Nobel Prize : an unexpected life in science /
J. Michael Bishop.
p. cm.—(The Jerusalem-Harvard lectures)
Includes bibliographical references and index.
ISBN 0-674-00880-4 (alk. paper)
1. Bishop, J. Michael, 1936– 2. Medical scientists—United States—Biography.
3. Oncogenes. 4. Nobel Prizes. I. Title. II. Series.

RC268.42.B57 2003
610'.92—dc21   2002192234
[B]

To the memory of
my mother and father

Contents

Illustrations

Preface

We live in an age defined by science, when many of nature's great puzzles have been solved. Despite transcendent achievement, however, science now finds itself in paradoxical strife with society: admired but mistrusted; offering hope for the future but creating ambiguous choice; richly supported yet unable to fulfill all its promise; boasting remarkable advances but criticized for not serving more directly the goals of society.

One of the contributing difficulties is that the general public does not understand who scientists are, what they do, or how they do it. Evidence for this appears regularly in films and television programs (and advertisements on television, in particular) that attempt to portray scientists. The caricatures on view there are often grotesque misrepresentations. Scientists are perceived as somehow not quite human.

The invitation to deliver three lectures in Jerusalem at the behest of Hebrew University and Harvard University Press offered me an opportunity to address the misapprehensions about scientists and their pursuits. I constructed a historical narrative that might reveal something about becoming a scientist and something about the practice of science. That narrative was the starting point for this book.

While I was writing, a colleague asked what my theme might be. I gave an answer that seemed trite, but was also truthful. I wrote this book to show that scientists are supremely human. *The Double Helix* by James D. Watson is one of the most widely read books about science and scientists ever published. Francis Crick offered an explanation for this success: "The layman is delighted to learn that after all, in spite of science being so impossibly difficult to understand, *Scientists Are Human*."[1] In the practice of science, we seek to understand ourselves and the world in which we dwell. If I have captured even some small part of this humane endeavor and made it generally accessible, I will consider the effort to have been worthwhile.

I wrote for the general public. I had hoped to avoid the use of note numbers scattered through the text, amounting as they do to "barbed wire keeping the reader at arm's length."[2] But my extensive use of quotations from other authors made notes unavoidable. In some instances, the notes serve simply to document the source of a quotation; in others, I have taken the opportunity to expand on the main text. Readers should pay no attention to any of this unless their curiosity is piqued.

The title of this book deliberately promises more than can be delivered. Although I was privileged to receive the Nobel Prize (along with my friend and erstwhile colleague, Harold Varmus), I have not written an instruction manual for pursuit of the prize, could not do so, and would not wish to do so. Rather, I have written with some ambivalence about the prize itself, and with some levity about the trappings that attend the prize. These are attitudes that probably come to mind more easily—perhaps too easily—once the prize is in hand. But make no mistake: I am in awe of the company in which I now find myself, and I am grateful to those who placed me in that company.

I had three principal companions in the stories I tell here. The first was my wife, Kathryn, who would prefer fewer of my apologies and more of my time. The second was Harold Varmus, whose partnership expanded my capabilities as a scientist and teacher. The third was Leon Levintow, an alter ego throughout my research career. Harold and Leon were daily presences in my professional life for many years. But Kathryn was the enduring presence behind the stage, the proverbial "astonished woman behind every successful man."[3] I also owe a large debt to Warren Levinson, who facilitated my embarking on the study of tumor viruses and remained a valued colleague in the years that followed.

The modern research laboratory can be a large and complicated social organism. At its peak, the research group that Harold and I directed together for more than a decade numbered at least two dozen. It was held together by the valiant efforts of several long-term staff. Of these, Joyce Futa, Jean Jackson, Suzanne Ortiz, Nancy Quintrell, and Lois Serxner deserve special notice. They have my enduring gratitude. I am also indebted to the many young scientists who trained with Harold and me, providing the intellectual yeast, the technical brawn, and the collegiality required for the success that we enjoyed.

I am grateful to Hebrew University and Harvard University Press for honoring me with the lectureship that sired this book. I thank President Menahem Magidor and Rector Menahem Ben-Sasson of Hebrew University; Dorothy Harman, representative of Harvard University Press in Jerusalem; and Professor Gideon Foerster, of Hebrew University, for their hospitality. I was especially moved by the opportunity to dine with the great Israeli poet Yehuda Amichai, whose poems I have long admired, and whose subsequent death has deprived us all of an impassioned voice for peace in the Middle East. The lectures were delivered in January 2000, a time when it was possible for Kathryn and me to walk through all of Jerusalem without fear.

I thank the following friends and colleagues, all of whom read the entire manuscript and none of whom minced any words: Bruce Alberts, Constance Casey, Julie Giacobassi, Zach Hall, Elizabeth Marincola, Susan Montrose, Miranda Robertson, and Harold Varmus. I offer the obligatory acknowledgment that all remaining imperfections are my responsibility. Sharon Carman helped in procuring the figures, and Grace Stauffer provided essential assistance with the manuscript. I thank Michael Fisher and Sara Davis at Harvard University Press for their patience and help as I recrafted the lectures into a publishable book, and Julie Carlson for skillful yet tolerant editing.

Science, that gives man hope to live without lies
Or blast himself off the earth; curb science
Until morality catches up?
But look:
At present morality is running rapidly retrograde,
You'd have to turn science, too, back to the witch doctors,
the myth drunkards. Besides that,
Morality is not an end in itself; truth is an end.
To seek the truth is better than good works, better than survival,
Holier than innocence, and higher than love.

—Robinson Jeffers, "Curb Science?"

The only solid piece of scientific truth about which I feel totally confident is that we are profoundly ignorant about nature. Indeed, I regard this as the major discovery of the past hundred years of biology. It is, in its way, an illuminating piece of news.

—Lewis Thomas, The Medusa and the Snail

CHAPTER 1

The Phone Call

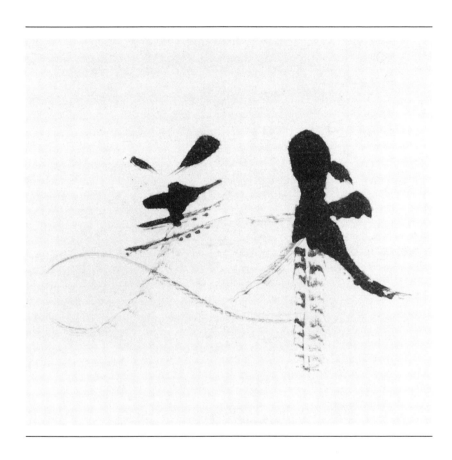

Fame is a fickle food
Upon a shifting plate . . .
Men eat of it and die.

—*Emily Dickinson*

Art by Tom Marioni, 1989. The Chinese symbol for "art" combines the character for "beauty" (left) with that for "skill" (right). The image was drawn with a seagull feather dipped in ink. (Reproduced by permission of the artist.)

At 3:00 A.M. on the morning of October 9, 1989, my older son, Dylan, took a phone call in his bedroom. My wife, Kathryn, and I had not heard the phone ring. As parents of two teenage boys, and as hostages to an irrational parsimony that permitted us only one phone line, we had long since inactivated the phone bell in our bedroom. Alert to our anxiety at being awakened so early in the morning, Dylan entered our room and announced quietly: "Don't worry Dad: it's NBC with good news." And good news it was, after a fashion. An announcement had just come from Stockholm that my colleague Harold Varmus and I would receive the Nobel Prize in Physiology or Medicine.

I spent the next hour answering calls from the press, doggedly cautioning all callers that I had received no notice from the Nobel Foundation and struggling through the mental haze of early morning to find similes that would make the prize-winning research accessible to the press and their readers. Then Kathryn and I took to the neighborhood streets and walked off the shock as dawn broke. Thus it was that the Nobel Foundation never reached me directly with the news. Instead, someone read the citation to Dylan over the phone in my absence. Having no experience with Scandinavian accents, Dylan understood not a word. A confirming telegram arrived a day later. Until then, an inner voice kept insisting that I was being made the butt of a gargantuan practical joke.

There was little joy for me in those dawn moments. Instead, I was disquieted by two opposing thoughts. On the one hand, I felt less than fully deserving, because the discovery for which Harold and I were being honored was only in modest part of my own making. On the other hand, I knew that this might not have been the first time for me, and the opportunity that I had squandered a few years before had been entirely of my own making (more of this in Chapter 2). I was also trou-

"Nobel Attire" by Jamie Simon,
1989. The author on the left,
Harold E. Varmus on the right.
(Reproduced by permission of the
artist.)

bled that I seemed to care—surely none of this would matter in the long view.

My family and I were not exactly ready for that call from Stockholm. I myself had a cosmic conflict. I was expected at an 11:00 A.M. press conference, but I also had tickets for a crucial playoff game between the Chicago Cubs and the San Francisco Giants baseball teams. Let the record show that I am an ardent Giants fan. I was unflinching: the press conference was moved to 8:30 A.M. so that I could arrive at the ballpark in time for batting practice, an essential ritual for the cognoscenti. The Giants won the game and, thus, the National League Championship when Will Clark drove a two-out single "up the middle" off a no-balls and two-strike pitch from Mitch Williams. I will remember that piece of trivia long after I have forgotten Avogadro's number (a physical constant of use to some scientists, but that does not come easily to my mind even now). And why not? Hitting a baseball from "behind in the count" is a supremely difficult endeavor, a metaphor for life.

Kathryn found it necessary to buy a new gown. She finally did, well after we had arrived in Stockholm, the day before the ceremonies—chutzpah of the first order. It must be said, however, that she was given spectacular attention in the store. In contrast to the United States, where celebratory summons to the White House are more frequent for championship sports teams than for scientists, Sweden honors Nobel laureates above all others. I provide a case in point. The schoolteachers of Stockholm were on strike during the days that Harold and I were in Stockholm to receive our laurels. They carried two sorts of placards on the picket lines: one protesting their salaries, the other apologizing to the Nobel laureates for distracting attention from the ceremonies.

Son Dylan was astonished when the prize was announced at his high school assembly, to cheers usually reserved for victorious sports heroes. "Why didn't you prepare us for this, Dad?" he asked. My only answer could be that I was not prepared myself, indeed, had not expected the occasion despite years of rumor and omen. The Nobel Prize had seemed remote through much of my prior life. I have no recollection that I knew of its existence until I arrived at Harvard Medical School. There the prize was never far from the communal consciousness, so I at least learned its name. But my own entry into science was so unlikely and so difficult (see Chapter 2) that achievement worthy of the Nobel remained beyond my wildest dreams as I climbed the academic ladder.

My younger son, Eliot, insisted on going to his middle school at 6:30 A.M. in order to sort basketball jerseys. Pleas that he might one day regret having missed this special morning at home went unheeded. On arrival at school, however, he was greeted by an excited member of the staff who happened to be Swedish and who forthwith sent him home to celebrate what he by now understood was something beyond the ordinary. So it was that our entire family was able to watch the Giants subdue the Cubs. (Dylan had excused himself from school of his own accord, demonstrating the will of late adolescence.)

It was Eliot who, in his innocence, kept matters in perspective that morning. Once out of the house and away from me, he asked my wife: "OK Mom, what is this Nobel stuff about, anyway?" Neither my wife nor I had a ready answer. But I have taken moments during the intervening years to reflect on what is admittedly a very good question.

Alfred Nobel

The answer to Eliot's question would have to begin with the man who provided the "stuff," Alfred Nobel.[1] This man and the unusual philanthropy that he left behind at his death created a benchmark for achievement that reigns supreme over both scientists and the general public alike. If my story is to be fully appreciated, his must first be told.

Born in Stockholm in 1833, Nobel lived for sixty-three years, achieving renown as a fabulously wealthy, lonely, and itinerant individual—newspapers of his time styled him as the "world's wealthiest tramp" (the French novelist Victor Hugo apparently originated the description). Alfred's father, Immanuel, was a minor industrialist whose financial status fluctuated alarmingly between affluence and bankruptcy. Immanuel was an inventor of sorts. He never achieved the experimental prowess of his son, Alfred, but late in life, he did invent plywood. Alfred's mother, Andriette, came from a family of some means, but endured the economic misfortunes of her husband without complaint and became "Alfred's universe."[2] Alfred remained deeply attached to his mother throughout her long life and, after her death at the age of eighty-nine, donated his portion of her estate to the Karolinska Institute in Stockholm to establish the "Caroline Andriette Nobel Fund for Medical Research"—a harbinger of greater largess to come.

On December 4, 1837, papa Immanuel fled Stockholm for Russia, to seek his fortune and avoid debtor's prison in Sweden. Andriette and the Nobel children would follow only five years later, after suffering great deprivation and surviving only because Andriette's father came to the rescue. The reunited family took up residence in St. Petersburg. There young Alfred lived a secluded youth, plagued by chronic ill health and educated by tutors at home, while his father dabbled with varying success in the munitions business. Alfred grew into an introverted and lonely adolescent with deep interests in chemistry, literature, and language that were to resonate throughout his life. He seems to have been permanently marked by the social disgrace that his family suffered during their period of poverty.

Alarmed by Alfred's reclusive personality, and hoping to divert him from his ambition to be a writer, his no-longer-impoverished parents sent him on an extended study tour of Europe and America when he was seventeen. One of Alfred's stops was Paris. There he struck up an enigmatic romance with a young Swedish woman who soon died of tuberculosis. The loss turned him into a "forsaken eremite in the world of the living" and moved him to morbid poetry.[3]

A second encounter in Paris, with a chemist named Ascanio Sobrero, proved more propitious. Three years before, Sobrero had learned how to combine glycerine, nitric acid, and sulfurous acid to make a substance he called nitroglycerine. Sobrero recognized his concoction as a powerful explosive, but had done nothing practical with it. Indeed, he had abandoned its study after realizing how easy it was to detonate the explosive accidentally—Sobrero himself suffered a severe facial injury in one laboratory accident with nitroglycerine.[4]

Young Alfred proved more determined. So in due course, the Nobel family set about to commercialize nitroglycerine, a highly unstable explosive that comes in the form of an oil and that was handled with astonishing lack of care in the Nobel establishment. Pictures from the time show the explosive liquid being carried by hand in open buckets. Eventually, an accidental explosion occurred at the Nobel factory in downtown Stockholm, killing one of Alfred's brothers. The plant was moved first to a suburb, and after further explosions, far from the city, where it remains today, now a model of industrial safety.

The tragedies brought out the inventor in Alfred. First, he devised a reliable detonator for nitroglycerine. (It was in fact the first detonator of any kind for explosives and gained Alfred considerable fame.) Then, in a tour de force, he invented a safer form of the explosive itself: the treacherous oil was impregnated into a solid base and dubbed "dynamite." These inventions brought Alfred to preeminence in the family business. He set out to consolidate and extend the manufacture of dynamite, establishing plants throughout Europe, and in the United States and South America.

As Alfred's fortune grew, so did his guilt. The explosives he first developed for civil engineering also transformed the conduct of war, and Alfred himself found munitions an endlessly fascinating subject. Seeking justification, Nobel conceived a prophetic rationalization: "The day when two contending armies can destroy each other within seconds, all civilized nations will retreat from war and demobilize their armies."[5] Nobel had anticipated the modern strategy of mutual deterrence. Ironically, nitroglycerine also found a medical use as the now familiar treatment for angina pectoris. Nobel himself suffered from this ailment and recognized the irony: "Isn't it the irony of fate that I have been prescribed [nitroglycerine], to be taken internally! They call it Trinitrin, so as not to scare the chemist and the public."[6]

Whatever its justification, Nobel's fascination with munitions eventually brought him grief. At the age of forty-three, he had established himself in Paris, a city that he loved passionately. While there, he invented a form of smokeless gunpowder that attracted the attention of military authorities throughout Europe. The Italians were the first to contract with Nobel for the development of his invention. The French took umbrage, eventually accusing Nobel of espionage and forcing the closure of his laboratory. Deeply disillusioned, Nobel left France and never returned.

While still in Paris, Alfred had wearied of his lonely life and had begun to search for a companion. He was known to enjoy the company of cultured and intelligent women, but none of these seemed to fully satisfy him: "I personally find the conversation of Parisians the dreariest thing I know, whereas it is delightful to meet cultured and not excessively emancipated Russian ladies. Unfortunately, they have an aversion to soap—but one must not expect too much."[7] Determined to find company, Alfred placed the nineteenth-century equivalent of a "personal ad" in Vienna newspapers, soliciting a live-in secretary. This was answered by one Bertha Kinsky (then age thirty-three to Alfred's forty-three). Bertha was rebounding from a romance with an Austrian count whose family had disapproved of her lesser lineage.

Alfred's expectations of companionship proved to be more intense than those of Bertha, so the arrangement lasted only a short while. Bertha soon eloped with her Austrian count, to become Frau von Suttner. But she and Nobel remained friends for life. She achieved international renown as a proponent of disarmament, encouraged Alfred to include peace among his plans for prizes to recognize great contributions to humanity, and eventually received the Nobel Peace

Prize herself, four years after its establishment (and nine after Alfred's death). Rumors of a sympathy vote persist to this day. Truth be told, however, Alfred was skeptical of Bertha's methods and influence: "Good intentions alone will not assure peace, nor, one might say, will great banquets and long speeches. You must have an acceptable plan to lay before governments. To demand disarmament is ridiculous and will gain nothing."[8] Alfred remained wedded to his faith in mutual deterrence by force of arms.

Rejected by Bertha, Alfred in the same year found Sophie Hess, a clerk in a flower shop on the outskirts of Vienna (age twenty to Alfred's forty-three). Theirs was not a conventional relationship for the times, which may explain why photographs of the two together or even Sophie alone are difficult to locate. They pursued a troubled relationship over fifteen years, with Alfred refusing marriage but Sophie nevertheless using his name in her personal affairs. Even Bertha von Suttner was misled, referring to Sophie as "Madame Nobel" in some of her correspondence with Alfred. Alfred paid his greatest compliment to Sophie when he took her to meet his elderly mother in Sweden. The encounter went surprisingly well.

Alfred could be harsh, particularly about Sophie's rough edges. He wrote to her in exasperation: "Ever since the first day, I asked you to get an essential education because it is not possible to really love someone who shames you daily through her lack of education and tact. Apparently, you are unaware of these flaws, otherwise you would at least have tried to smooth out the rough edges long ago. Even if one were head over heels in love, a letter such as you write would be a cold shower."[9] As the years passed, Nobel grew ever more resentful of his attachment to Sophie: "For many years I have sacrificed my time, my reputation, all my associations with the educated world and finally my business—all for a self-indulgent child who is not even capable of discerning the selflessness of those acts."[10]

Having failed with Sophie and not inclined to temper his disposition, Alfred lived out his life alone, settling eventually at San Remo on the Italian Riviera. There he built a laboratory and a rocket range to pursue his burgeoning interest in ballistic missiles. Nobel put the laboratory to good use. By the time he died, he held more than 350 patents. The rocket range was another matter. It extended out over the Mediterranean Sea, so Alfred's primitive missiles occasionally dropped among fleets of pleasure yachts (harmlessly, by all accounts).

Alfred died in Italy on December 10, 1896, in bleak circumstances that he had long predicted: alone except for servants and a physician. Nobel had also harbored a morbid fear of being buried alive. He acted on the fear with a specification at the end of his will that, "after my death, my arteries shall be cut open, . . . and death confirmed by a competent physician,"[11] adding for good measure that he wanted to be cremated (on another occasion, he suggested immersion of his body in sulfuric acid)—all of which displayed the punctiliousness that underlay his success as an industrialist. Only after Nobel's death did an astonished world learn that this enigmatic celebrity had bequeathed virtually his entire fortune to establish the Nobel Prizes, now awarded in a ceremony held each year on the anniversary of Alfred's death.

The Nobel Prizes

The bequest for the Nobel Prizes was spelled out in a single handwritten paragraph that named physics, chemistry, physiology or medicine, literature, and peace, in that order, as themes for the prizes. A prize for work in economics was established by the Bank of Sweden many years later (1968), in celebration of the Bank's three hundredth anniversary. The gesture caused great consternation among the Swedish stewards of the Nobel Prize, who saw it as an effort by a "non-rigorous discipline to cloak itself in the trappings of an objectivity it did not and could not possess."[12] To this day, the prize in economics is known as the "Bank of Sweden Prize in Economic Sciences in Memory of Alfred Nobel" to distinguish it from the "real" Nobels, and is administered by the Nobel Foundation but not paid out of the Nobel endowment. It is nevertheless universally perceived as a Nobel Prize, to quiet acquiescence by the Nobel Foundation and the "authentic" laureates (with the exception of an occasional physicist who voices a complaint).[13]

Nobel himself never accepted economics as a science, and even some of the laureates in economics have expressed doubt about the

Billboard near Gettysburg, Pennsylvania, October 1989. (From the family album of the author.)

prize. In protest against the award to the outspoken and controversial Milton Friedman in 1975, a previous economics laureate, Gunnar Myrdal, wrote an open letter to a Swedish newspaper calling for an end to the economics prize. Myrdal's colaureate (and ideological opponent), Friederich Hayek, toasted the king and queen of Sweden with the remark that he would have recommended against establishing the prize in economics had he been asked—in his view, the discipline was not sufficiently rigorous and objective. One authority on Alfred Nobel and his prizes has suggested that too many of the "Nobelized achievements" in economics "seem perilously close to scientizing the commonsensical."[14]

The order in which Nobel named the themes determines the order in which the prizes for science are handed over at the ceremony and the order in which the recipients parade into the banquet that follows. There is no indication that the order actually reflected a ranking of merit by Nobel, although physicists would probably like to believe otherwise. The Peace Prize is presented separately in Oslo, in a gesture that Nobel hoped would diminish the dyspeptic rivalry between Norway and Sweden.

The tangible symbols of the Nobel Prize take three forms: cash, medal, and certificate (or "diploma," as the Nobel Foundation calls it—one wag has suggested to me that this particular diploma symbolizes graduation into superannuation). The monetary awards have always been large by the standard of the times, which perhaps accounts for how quickly the Nobel Prize gained its international celebrity and its enduring supremacy over any other award for creative achievement.[15] This crass reckoning seems not to trouble most laureates. The growth in value has been particularly robust in recent years, inspiring envy among previous generations of laureates (who only begrudgingly make an allowance for inflation, a calculation that shows little real growth over the original value of the prize). U.S. scientists are further aggrieved because their nation is one of the few that taxes the Nobel Prize as income. This egalitarian practice began only in 1986— Harold and I were just three years too late to escape the tax collector. The combined efforts of the United States and the State of California claimed exactly half my share of the Nobel cash.

The medals are struck from gold. The laureates are also allowed to purchase three replicas of lesser value, which they can distribute as they see fit. I donated my replicas to the three institutions of higher education where I had studied (Gettysburg College, Harvard Medical School, and the University of California, San Francisco), where they have met diverse fates: Gettysburg College displays its copy at the entrance to the campus library, UCSF has yet to give its a permanent residence, and Harvard has presumably added its to a drawer bulging with others (ego forbids my asking). The medals bear on one side a visage of Alfred Nobel, on the other side an image symbolizing the subject of the award, along with the name of the recipient in letters small enough to be humbling. Here again, the economists have been set apart: their names appear only on the rim of the medal.

The fate of Nobel medals constitutes a small study in human foibles. They have been sold, lost, stolen, and fought over by relatives and heirs. Two Nobel medals stored in Niels Bohr's research institute in

Copenhagen were dissolved in strong acid to keep them from Nazi raiders in December of 1943. As soon as Denmark had been liberated from German occupation, the gold was recovered from the acid, with the medals recast by the Nobel Foundation and returned to their rightful owners.[16]

The diploma specifying the award in physiology or medicine brought my only disappointment with the pomp and circumstance in Stockholm, because it differs from all the others by not being decorated with an original work of graphic art. I find this tradition puzzling. Biologists are surely as receptive to the fine arts as any other sort of scientist. Indeed, my most vivid memory from Stockholm recalls a moment of epiphany in the presence of art. As one of the many small privileges extended to laureates, my brother, Stephen, and I had been admitted to the Thiels Gallery while it was officially closed and, thus, empty. Standing alone in a small room on the top floor, in the company of prints by Edward Munch and a death mask of Friedrich Nietzsche, I looked out a small window across snow-covered lawns and, for the first time, fully apprehended the turn my life had taken. It was a moment that only death will erase, and I doubt that it could have happened for me in any other setting.

I have encountered two widespread misconceptions regarding the symbolic manifestations of the Nobel Prize. The first is that the word "peace" is part of the title for all the prizes. Thus, recipients are said to have received the "Nobel Peace Prize for Physics, or Chemistry, or Physiology or Medicine." Curiously, I have never heard this error committed in reference to the literature prize. In my own instance, not only is peace added erroneously to the title, the word physiology is often omitted, reflecting a decline in the cachet of this classical discipline. The second misconception is that the medals are worn from a ribbon about the neck (evoking for me an image of the biblical millstone). This is not so. The medals contain no provision for attaching a ribbon, but are apparently intended to rest forever in the velvet-lined case in which they are received from the hands of the Swedish monarch. I like that scheme, because it helps preserve the luster of the medals, and because it discourages ostentation by making it inconvenient.

Pomp and Protocol

It would be difficult to exceed the pomp and circumstance of the No-bel ceremonies and its attendant events.[17] Royalty dominate the stage at the ceremony and the central table at the banquet that follows, fos-tering suggestions that the Swedes have retained a monarch in their social democracy mainly for the purposes of what is known as "Nobel Week" in Stockholm. The Swedish royalty learned the value of Nobel Prizes early. King Oskar II declined to present the first prizes in 1901—he had been skeptical about the prizes on several counts and advised Nobel's nephew, Emanuel: "It is your duty to your family to make sure that their interests are not jeopardized by your uncle's nonsensical ideas."[18] But the press coverage of the event was so extravagant that Oskar showed up the following year and reliably thereafter, and so have all of his successors. King Oskar expressed special disdain for the prize in peace, telling Emanuel, "Your uncle was talked into this by fa-natics, womenfolk mostly."[19] At least one woman certainly had some-thing to do with it—recall Bertha von Suttner—but she was hardly a "fanatic."

The ceremonies are embellished with a full symphony orchestra and distinguished vocalists. The flowers that bedeck the ceremonial hall are brought to Stockholm from Nobel's estate at San Remo, freshly harvested for the occasion. The banquet for more than one thousand guests is held in an immense and fabled hall within the city hall of Stockholm. The place settings are so valuable that they must be cleaned by hand—an exercise alleged to require more than a month's time. The guests of honor must all descend a long and treacherous stone staircase while trumpets blare and all the other guests look on, already seated at their tables. These are harrowing minutes. The women who must make the descent in formal attire receive prior in-struction on how to avoid a disabling fall, and the men escorting them have been taught how to prevent a fall by their partner without com-mitting an indiscretion.

Protocol rules with an iron hand. The laureates are taught how to bow after receiving their award from the hands of the king, coached in the Scandinavian toast known as the "skoal" (a maneuver that includes an impishly intimate exchange of glances when performed expertly), and told to pay more attention at dinners to whomever is seated to their right than to their left. Men must wear full evening dress for the ceremony and banquet, whereas women have their usual greater liberty with fashion. The Nobel Foundation assigns each laureate an aide-de-camp, often drawn from the Swedish diplomatic corps. They are available to assist on all occasions, but their principal purpose is to be certain that the laureates have donned tie and tails properly. My aide seemed singularly doubtful that I could meet this challenge unassisted, and he was correct.

The service of each course at the banquet is carefully choreographed. The placement of guests is prescribed by tradition. My wife, Kathryn, was seated to the left of the king and found herself in conversation about hunting and cross-country skiing, neither great passions of hers. I dined with the wife of the speaker of the Swedish Parliament, who could not be lured into a conversation about politics. A gala dance follows the banquet, but the laureates spend much of that time waiting for a brief private audience with the king and queen. Kathryn and I never did reach the dance floor, in part because we were due next at a party with Swedish medical students that would finish as dawn broke (and dawn breaks very late in December in Stockholm). We have been back to the ceremonies twice during the ensuing years, and we have made a point of dancing.

I recall vividly that dawn the day after. I awoke after four hours of sleep, at 10:30 A.M. The sun was barely above the horizon, shining into my eyes through one of the great front windows of the Grand Hotel. I was morose. The culminating moment had come and gone. The Nobel medal was in hand, offering no prospect other than the problem of storage. The transience of it all was oppressive. All that remained, I thought, was an inexorable decline into age, without reprieve from my personal demons.

The night following the ceremony and grand banquet, the king and queen host the laureates for dinner in the royal palace. The one hundred or so guests all sit at a single table of astonishing length, laden with gargantuan candelabra and elaborate silver dining ware that once belonged to some South American royalty. The company changes de-

The Bishop family with the King and Queen of Sweden, December 10, 1989.
(© The Nobel Foundation.)

cisively from the evening before. I was seated between two women of consequence, one Norwegian, the other Swedish, each of whom muttered invective about the other into my ear—it seemed that they were rivals in the diplomatic community. My wife fared better. She befriended the king's stablemaster, who later gave her a private tour of the royal stables (which did cater to a passion of hers) and offered her a ride in the royal sleigh (she never found the time for that).

The Swedes spare no effort to keep events flowing. I arrived in Stockholm with an upper respiratory infection that rapidly worsened. By the day before I was to deliver my Nobel Lecture, I had no voice whatsoever. My aide-de-camp quickly arranged a medical consultation with the otolaryngologist who cared for the soloists of the Stockholm Opera Company. (I had heard them sing Tosca the evening before.) At the appointed hour, the specialist happened to be on service at an immense psychiatric hospital on the outskirts of Stockholm. I was bundled off to the institution, past hordes of mercifully inattentive patients, to be treated with an unorthodox combination of antibiotic and steroid. Twenty-four hours later, I had recovered sufficiently to struggle through my forty-minute presentation, lubricated by copious amounts of mineral water—the glass is prominent in photographs of the event.

The laureates are allowed to bring their immediate family and twelve additional guests to the ceremonies and banquet. Some laureates manage to expand their allotment of guests by artful brokering with other laureates and acquaintances among the Swedish scientific elite. The current record is said to exceed two dozen. But the laureate in literature for 1989, Camilo José Cela, trumped even that record by insisting on bringing with him fifty residents of the small Spanish town in which he had been born. They were not accommodated in either the ceremonial or banquet hall, but they were able to watch all of the proceedings on closed circuit television and dined separately in a private room.

Cela's reputation as an iconoclast had preceded him to Stockholm. He did not disappoint. He brought with him not his wife (whom he later divorced), but a younger and glamorous "muse" (whom he later married). The "muse" doubled as business agent—she was reputed to be charging for his interviews with the press (a sore point with the scientist laureates, who could not have commanded a krona for interviews). Cela earned the Nobel citation with his fiction, but he was also a telejournalist, renowned as the first individual to use profanity on Spanish television and for publishing a lexicon of unsavory words and expressions that were widely used by the public but never by the press. His iconoclasm provoked me to read his novels. I did and I was pleased. Cela died just as I was completing this text. His obituary may have been even better reading than his fiction.

Others have broken the grip of protocol in more subtle ways. Among these renegades was the late Howard Temin, a friend and Nobel laureate whose work helped set the stage for Harold's and mine. At the banquet, representative laureates deliver brief remarks to the glittering and not especially attentive audience. The remarks are usually laced with gratitude and revelry. But Howard used his moments at the microphone in 1975 to berate the tobacco industry for its baleful effects on human health, and his audience for making liberal use of tobacco products even as he spoke. It is a tragic irony that Howard later died prematurely of lung cancer, albeit not a form attributable to smoking. By 1989, smoking had been banned from the banquet hall (although, to this day, fine cigars—an oxymoron for some—remain in evidence following the dinner at the palace).

Protocol dictates that once refused, the Nobel Prize cannot be reclaimed. Two Nobelists have formally refused the prize (both in literature): Boris Pasternak, in 1958 under compulsion from the government of the Soviet Union; and Jean Paul Sartre, in 1964 for reasons of his own. Gerhard Domagk, Richard Kuhn, and Adolf Butenandt were all German scientists who were named to receive the Nobel Prize and were glad of that (Domagk for physiology or medicine in 1939, the other two for chemistry in 1938 and 1939, respectively), but were prohibited from attending the ceremonies by Adolf Hitler. The Nobel Foundation had previously offended Hitler by awarding the Peace Prize to the German journalist Carl von Ossietzky, a militant pacifist and anti-Nazi who was in a prison hospital when his Nobel award was announced in 1935. Ossietzky remained imprisoned and died two years later of tuberculosis, without receiving the symbols or the financial benefits of the award. Domagk, Kuhn, and Butenandt were given their medals, but for reasons never specified, not the money, after the conclusion of the Second World War.

Harold and I saw the protocol governing refusals pleasingly bent. The Soviet government had compelled Boris Pasternak to decline the Nobel Prize in 1958 because it was viewed as a reward for the dissident political innuendo in his novel, *Doctor Zhivago*. Pasternak died without ever receiving any token of his honor. At a reception held the day before the prize ceremony in 1989, while Harold, I, and our fellow laureates looked on, the Nobel Foundation finally paid formal tribute to Pasternak by presenting the medal intended for him to his son, Evgenji (who wryly asked whether the monetary award was also forthcoming—it was not).

Learning of this special tribute, and citing his close friendship with Pasternak, the renowned Russian cellist Mstislav Rostropovitch offered to make a few remarks at the banquet in memory of his friend. The Nobel Foundation politely said no, only laureates speak at the banquet. Rostropovitch countered by offering to play his cello. That was an offer not even the Nobel authorities could refuse. So the banquet concluded with a Bach suite, providing the most poignant moments of my days in Stockholm.

Legacies

At his death, Nobel was widely renowned for his wealth and for his skills as an industrialist and inventor. But despite all of his accomplishments, Alfred took a dim view of himself: "I drift about without rudder or compass, a wreck on the sea of life; I have no memories to cheer me, no pleasant illusions of the future to comfort me, or about myself to satisfy my vanity. I have no family to furnish the only kind of survival that concerns us, no friends for the wholesome development of my affections, or enemies for my malice."[20]

Why did this lonely and discontented curmudgeon establish the prizes that now bear his name? The answer includes guilt and a quest for redemption. Nobel's brother Ludwig died in 1888, when Alfred was fifty-five years of age. Some of the press mistakenly believed that Alfred himself had died and published obituaries that described him as a merchant of death. Moved to obsession with his posthumous reputation, Nobel rewrote his will to create a legacy that everyone could honor. The new will also helped Alfred with a dilemma over the disposal of his fortune. In his own words:

> I regard large inherited wealth as a misfortune which merely serves to dull men's faculties. A man who possesses great wealth should therefore allow only a small portion to descend to his relatives. Even if he have [sic] children I consider it a mistake to hand over to them considerable sums of money beyond what is necessary for their education. To do so merely encourages laziness, and impedes the healthy development of the individual's capacity to make an independent position for himself.[21]

I once read that passage to the affluent student body of a private high school. The students were not amused and I have not repeated the performance.

True to his word, Alfred left only a small fraction of his fortune to relatives, reserving the balance for the prizes. The entire legacy was valued at 33 million kronor (more than $200 million in today's currency), of which no heir received more than 300,000 kronor. A modest annual income was also provided for Sophie Hess. Alfred's extended family immediately asked that the will be declared invalid. The dispute was resolved only after three years of diligent adjudication by Nobel's deputy, a twenty-five-year-old chemical engineer named Ragnar Sohlman, who used wily subterfuges to shelter the estate from the tax authorities while bringing the legatees to terms.

The negotiations with Sofie Hess were especially distasteful. Sofie was dissatisfied with the size of her annuity and threatened to make public the more than two hundred letters that she had received from Alfred, some of which were decidedly indiscrete by the standard of the times. Sohlman bought the letters from her in return for a permanent injunction against any public comment by Sofie about her relationship with Alfred.[22]

The French were particularly difficult about taxes. Having driven Nobel from the country years before, they now decided that he had nevertheless been a permanent resident there and attempted to tax his estate accordingly. Sohlman foiled them, systematically moving all of the securities that Nobel had stashed in Paris to England and Sweden, where they could be safely liquidated. The transfer required that he personally accompany the securities to the Swedish consulate, and then on to a railway station, revolver in hand to defend against possible theft. His legacy remains in the person of his grandson, Michael, who currently serves as executive director of the Nobel Foundation and presides over an endowment equivalent to more than $430 million. Aspirants to become Nobel laureates need not fear for the monetary promise of the prize.

Nobel left one other legacy: a play entitled *Nemesis*, which he published privately. A copy survives in the archives of the Nobel Foundation in Stockholm. I have not read the play, but it is said to be dreadful—Nobel's family tried to have all copies destroyed after he died.

The Nobel Decision

Nobel specified that his intent was to honor individuals whose work had given "the most benefit to humankind in the preceding year."[23]

This stipulation of promptitude proved impossible to meet, and in fact probably contributed to some embarrassing mistakes, such as the Nobel Prize awarded in 1926 to Johannes Fibiger, a Danish scientist who claimed that worms cause stomach cancer (they do not), or the one in 1949 to Antonio Egas Moniz, a Portugese neuroscientist who introduced prefrontal lobotomy as therapy for psychiatric disorders (a tragic misapprehension now discredited). It generally takes some years for the full significance of a discovery to become apparent. So most of the prizes are now given for work performed a decade or more previously. Clever lawyers can usually break a will.

The Nobel Foundation solicits nominations from more than a thousand individuals and institutions, then refers the nominees to individual committees for each of the prizes. The many individuals who submit unsolicited nominations are engaging in futility, but nevertheless account for many of the publicized claims, "nominated for the Nobel Prize." The responsibility for awarding the prizes in physics, chemistry, and economics lies with the Royal Swedish Academy of Sciences; for the prize in physiology or medicine, with the Karolinska Institute in Stockholm; for the prize in literature, with the Swedish Academy; and for the prize in peace, with the Norwegian Parliament. The diversity of the responsible bodies helps account for idiosyncratic variations in the quality of the laureates. In particular, the committees that elect the laureates in literature and peace are sometimes accused of allowing various ancillary considerations to influence their decisions unduly.

The nominations are studied for six months and the recorded deliberations held in confidence for fifty years. (As an officer of a public institution that is subject to U.S. "sunshine laws," I would welcome the luxury of occasionally conducting our affairs in such confidence.) The fifty-year-old archives were first opened to public scrutiny in 1975. Historians of science have been enjoying a field day ever since, exploring the machinations that sometimes underlie the choice of laureates. The most celebrated case has been that of Albert Einstein. His nomination for the theory of special relativity was resisted strenuously by an influential member of the award committee and others in the Swedish scientific community, who found "theoretical science" to be inappropriate for a Nobel and displayed a hidebound refusal to accept the staggering implications of relativity. The matter became so contentious that the Royal Swedish Academy of Sciences deferred the 1921 prize in physics following a stalemate in the debate over Einstein.[24]

The stalemate was broken in 1922, with the decision to honor Einstein for another piece of pathbreaking work—his explanation of how light can cause certain metals to emit electrons (the "photoelectric effect"). So it was that in December of 1922 Einstein received the 1921 Nobel Prize in Physics. Einstein was visiting Japan at the time and did not attend the ceremonies. He was represented by the German envoy, Rudolf Nodolny, in the face of ongoing confusion over Einstein's citizenship—both Germany and Switzerland were eager to claim him. We have no record of where or when Einstein received official notice of the award; even his travel diary from the journey does not mention the moment.

It is generally agreed that Einstein's work on the photoelectric effect was itself prizeworthy (albeit also theoretical). It correctly proposed that light is transmitted as discrete units known as "quanta," and it played a vital role in establishing the quantum theory that now underlies much of modern physics. Einstein himself considered his light-quanta proposal more revolutionary than his theory of special relativity. But still, the photoelectric effect served as a surrogate for relativity in honoring Einstein with a Nobel Prize. It was the theory of relativity that made Einstein first among equals in the world of physics, and gave him a celebrity that remains unique among scientists. The Nobel citation hinted at what had transpired: "To Albert Einstein, for his services to theoretical physics and especially for his discovery of the law of the photoelectric effect." The secretary of the Swedish Academy of Science drove the point home by writing to Einstein that his prize had been given "without taking into account the value which will be accorded your relativity and gravitation theories after these are confirmed in the future."[25] A second prize, seemingly in order, never came.

Machinations complete, the Nobel committees eventually report to the full membership of the responsible institution, which then votes on the recommendations. Affirmation is customary, but there have been occasional reversals or revisions of committee reports. In particular, there have been recurrent disputes over the merits of applied as opposed to fundamental research, with advocates of applied research usually finding themselves in the minority (despite the expressed purpose of Nobel to honor those who have "conferred the greatest benefit on mankind").

The preference of Nobel committees for fundamental over applied research sometimes puzzles the general public. There is no better example of this than the vaccines against poliovirus. Jonas Salk developed the first such vaccine, using virus that had been killed with chemicals, and became a medical icon, renowned as the "man who gave summer back to children." Until his death in 1995, Salk's authority with the public on any issue of medical science was astonishing. Many expected him to produce a cure for AIDS. He tried, but failed. Albert Sabin developed a different sort of poliovaccine, in which the virus remains alive but usually harmless. It took Sabin longer than Salk to succeed, but he lived to see his vaccine used throughout the world to take poliovirus to the brink of extinction. With the job largely done, the United States has now reversed course and switched to an improved version of the Salk vaccine, because Sabin's version occasionally regains the ability to cause disease.

The entire civilized world expected Salk, and perhaps Sabin, to receive the Nobel Prize in Medicine or Physiology. Neither did. The prize for poliovaccine went to John Enders, Thomas Weller, and Frederick Robbins, who had discovered that poliovirus could be propagated and enumerated in laboratory preparations of monkey cells. The Nobel committee rightfully recognized that this advance was essential to the subsequent development, evaluation, and production of the vaccines, and thus represented the fundamental step that made possible the elimination of poliovirus and the dread disease that it causes. Most medical scientists now agree that the Swedes made the right choice.

No more than three individuals can receive the prize in each category. The Peace Prize is unusual in occasionally being awarded to an organization instead of, or in addition to, individuals (the award to the United Nations and Kofi Annan is a recent example). Dividing the prizes is far more common in the sciences than in literature or peace, reflecting the fact that most discoveries in modern science arise from the efforts of multiple individuals. Publications in particle physics may have several hundred authors, and biology is moving into the same realm with the advent of the industrial-scale work required for sequencing genomes.

Thus, as the number of individuals responsible for single breakthroughs in scientific research has gradually increased, so too has the sentiment that a limit of three recipients for each prize may be too restrictive. The current limit of three for each prize is itself a compromise, representing a revision of Nobel's original bequest, which specified only one recipient per prize. Might the Nobel Foundation now be tempted to make the awards even more inclusive? It appears not. "There are no plans to change the rule," according to the current chair of the board for the foundation.[26]

Nobel prizes are not awarded posthumously. The public is not privy to whether a life-threatening illness might accelerate the anointment of a deserving candidate, nor to whether the premature death of one member of a research team affects the chances of the other members. There is no doubt, however, that survivors can on occasion prevail. Marie Curie is the supreme example.[27] She and her husband, Pierre, received the third Nobel Prize in Physics in 1903, for their contributions to the discovery of radioactivity (it appears that Marie coined the term). Marie's prize came just six months after she had been granted her doctoral degree. In 1906, Pierre was run over by a carriage in a Parisian street and killed. The grief-stricken Marie picked up their work and carried on, clothed in black for the remainder of her life. By 1911, she had received a second Nobel Prize, this one in chemistry, for the discovery of radium and polonium.

Marie Curie was an astonishing individual. She conducted her research under deplorable conditions, sustained radiation damage that eventually killed her, and racked up a series of memorable firsts—the first woman to earn a doctorate in France, the first woman to hold a professorship at the Sorbonne (indeed, anywhere in French higher education), the first woman to receive the Nobel Prize (preceding Nobel's erstwhile companion, Bertha von Suttner, by two years),[28] the first person to receive two Nobel Prizes (there have been only three such since),[29] the first woman admitted to perpetual interment at the Pantheon in Paris, and the first person to threaten the laureatehood with scandal when, as a widow, she entered into a highly publicized liaison with the renowned (and married) physicist Paul Langevin (once a student of her husband). Albert Einstein once accused Marie of having "the soul of a herring."[30] That now hardly seems credible. And there can be no doubt that her remarkable and celebrated life gave an early boost to the cachet of the Nobel Prize.[31]

In some instances, death may simplify the task of the Nobel committees. A prominent example concerns Rosalind Franklin, who pro-

"It was just that one time that you won the Nobel Prize, wasn't it, dear?"

duced much of the data required to solve the three-dimensional structure of DNA—a discovery that ranks among the greatest in all the history of biology.[32] Franklin died of cancer before the Nobel Prize for this discovery was eventually awarded to Francis Crick, James Watson, and Maurice Wilkins (who was Franklin's collaborator in what was anything but a collegial relationship).

If Rosalind Franklin had not died, how would the Nobel committee for the prize in physiology or medicine have met its obligation to choose only three recipients? Would the committee have omitted both Franklin and Wilkins? (It was Crick and Watson whose inspired interpretation of the data solved the puzzle.) Would Franklin alone have been omitted? (Those who believe that Franklin was the victim of gender bias both during her career and posthumously would expect so.) Might two Nobel committees have divided the spoils? (Prizes in chemistry and physiology or medicine could probably have been arranged to accommodate the foursome.) Our first chance for answers to these questions may come in the year 2012, when the files for the DNA prize can be opened to public scrutiny.

The criterion for selection in the sciences is explicit and strict: the laureate must have made a seminal discovery, preferably embodied in a single publication. Some of the discontent over selection of the laureates arises from confusion between discovery and accomplishment.

A scientist may make a single important discovery during a career without being otherwise very productive. On occasion, such scientists are honored with the Nobel Prize—true to the standard prescribed by Nobel, but annoying to the scientific community, which tends to denigrate the "flash in the pan." As Max Perutz, winner of the 1962 Nobel Prize in Chemistry, once wrote, "Success in research is a haphazard business, and great discoveries are not always made by great thinkers. Some are made by skilled craftsmen, some by observant watchmen, and some even by prosaic people doing a regular job because they are paid for it."[33]

In contrast, some scientists display remarkable industry, even virtuosity, without ever happening upon a discovery that substantially alters the course of science. Prodigious energy can earn a scientist achievement and celebrity, but these are not in themselves touchstones for Nobel committees.

Invention offers yet another set of challenges to Nobel committees. On some occasions, invention can involve genuine discovery, on others, it is based on previously established facts and principles. Inventions of both forms have been honored by the Nobel Prize for their great benefit to science or human welfare. But on the whole, discovery rather than invention rules (in Stockholm, at least; otherwise, invention often carries far more lucrative rewards). Had the Nobel committee been confronted with the example of nitroglycerine, they would probably have chosen Sobrero over Nobel, the creator over the exploiter. Yet it was the exploiter who realized great fortune and lasting fame.

The criteria for the laureates in literature and peace are necessarily more subjective. Nobel specified that the literature prize should recognize the "most outstanding work of an idealistic tendency"—hardly a formula for objective choice.[34] So it is not surprising that the Nobel Prizes in literature are more often controversial than those in the natural sciences. How could the judges for literature have ignored the likes of Leo Tolstoy, James Joyce, Willa Cather, and Virginia Woolf? But they did. In contrast, few of even the most assiduous readers are familiar with Par Lagerkvist, Frans Sillanpaa, or Henrik Pontoppidan—all Nobel laureates in literature, and perhaps not coincidentally, all Scandinavian. If the apparent bias was real, the fault did not lie with Alfred Nobel, who clearly specified the international nature of his intent: "It is my express wish that in awarding the prizes no consideration be given to the nationality of the candidates, so that the most worthy shall receive the prize, whether he [sic] be a Scandinavian or not."[35]

As for economics, only its immediate practitioners seem capable of appreciating the merits of its Nobelists. One perennial joke is that mere membership on the faculty of economics at the University of Chicago is sufficient to procure a Nobel Prize. Another is that although the prize for economics was instituted only in 1969, the field of eligible candidates may already have been exhausted. One administrator of the prize has told the press that "all the mighty firs have fallen; now there are only bushes left."[36]

In search of mighty firs, the committee for the prize in economics sometimes turns to mathematics, a common tool for modern economists. A notable contemporary example is the mathematician John Nash, who received the 1994 prize in economics for his work on game theory. The work was performed while Nash was still a student, nearly half a century earlier. In the interim, Nash developed schizophrenia and was largely unproductive. These circumstances caused great controversy within the Nobel committee and the prize to Nash came very close to being scuttled in the final balloting by the Swedish Academy of Science. The details of this controversy are known only because of an unusual and extensive breach of the secrecy that normally envelops the Nobel deliberations.

Afterward, a committee of the academy recommended that the purview of the economics prize be broadened to include all social sciences. The recommendation has never been acted upon formally. Nash has achieved considerable celebrity with the general public through the film and biography of the same name, *A Beautiful Mind*.[37] His example calls to mind the venerable myth that Nobel excluded mathematics from his prizes because his wife had been unfaithful to him with a mathematician. There is no substance to this myth, particularly since Nobel never married.

On the whole, Nobel laureates represent a humbling array of dedicated talent. Joining a list that includes Albert Camus, Francis Crick, four Curies, Albert Einstein, T. S. Eliot, William Faulkner, Boris Pasternak, and James Watson puts a distinct perspective on life, not a perspective that is necessarily easy to live with. How could I belong in that company?

The final decisions are announced in early October. It is rumored that potential laureates spend the appropriate October night lying awake, anxiously awaiting a phone call from Stockholm. But I for one went to sleep on that fateful night oblivious of what might come in the hours ahead, and the same seems to have been true of many laureates before and after me. I confess, however, that when Dylan woke me early the next morning for that phone call from NBC, my immediate thought was, "Good Lord, this is the second week in October." Apparently, the matter had not been completely out of mind.

Onus and the Nobel Prize

The Nobel Prize is not an unmitigated blessing.[38] The comic strip *Doonesbury* once portrayed an alumnus at a college reunion who expressed regret over his drab life as a chemist. When asked whether the fact that he had received the Nobel Prize had not made him feel better, the chemist responded: "Not really. I just could not relate to it." That conversation almost certainly took place in California.

The great astronomer Subramanyan Chandrasekhar has also spoken adversely of his Nobel Prize: "This thing [makes] a huge perturbation in my life [and] is not something which I have particularly liked . . . in many ways I would have much preferred not to have received it . . . it is well to remember that there is in general no correlation between the judgment of posterity and the judgment of contemporaries."[39] Chandrasekhar went on to express discontent with his life

Potential Surface of Density by Terry Winters, 1996. (Reproduced by permission of the artist.)

despite his extraordinarily successful career, regret that he had imposed his obsessional lifestyle on his wife, and chagrin that his life has been so "one-sided," so "lonely," so "inescapable." It is sobering to hear such discontent from a Nobel laureate whose achievements are so celebrated that his name graces a space satellite.

I too must say that although receiving the Nobel Prize was a surreal experience that I welcomed, it has not enriched my life by any large measure, has not changed the way I feel about myself, and has not changed the way my colleagues feel about me—they know me too well to be swayed by a single phone call from Stockholm. There are other perceptions, of course, but most are erroneous.

It is commonly said, for example, that Nobel laureates are instantaneously regarded as universal experts, asked to comment on virtually every aspect of human existence—crime, poverty, foreign policy, religion, and the future of humankind. More than a decade after the announcement of my own award, however, I have yet to be asked about any of these things (not a flattering admission, I realize). It was otherwise, however, for Albert Einstein. During a trip to New York City in 1930, he was asked, "within one brief quarter of an hour, to define the fourth dimension in one word, state his theory of relativity in one sentence, give his views on prohibition, comment on politics and religion, and discuss the virtues of his violin."[40]

It is also widely believed that receiving a Nobel Prize in science opens a cornucopia of research funding and a gateway to unimpeded publication. The reporters who questioned Harold and me during our press conference about the prize insisted that this must be true. Again, I regret to report otherwise (another unflattering admission).

Still, the Nobel Prize does carry a measure of brief celebrity, those fifteen minutes prescribed by Andy Warhol. So it was for Harold and me, almost to the minute. Since I had made my point about baseball on that fateful first day, it came to pass that Harold and I were asked to throw out the first pitch at the fourth game of the 1989 World Series, which matched the San Francisco Giants against the Oakland Athletics. Our performances on the mound would be broadcast on nationwide television.

Those performances never came. The devastating Loma Prieta

earthquake of October 17, 1989, intervened, killing several dozen individuals, leveling more than a mile of elevated freeway, dislodging one section of the bridge that spans the San Francisco Bay between San Francisco and Oakland, destroying several blocks of homes in San Francisco, and disrupting the schedule for the World Series.

I was in the baseball stadium, awaiting the beginning of the third game, when the earthquake struck. The magnitude of the quake was not immediately apparent to those of us in the lower deck of the stadium. But the hasty departure of the Goodyear blimp from above the stadium suggested that something momentous was afoot, and I had the fleeting intuition that true celebrity would now elude me. That intuition proved prescient.

The original plan was for the iconic Willie Mays to throw out the first pitch at the third game, Harold and I at the fourth. After the earthquake, Willie's place at the third game was taken by twenty "heroes of the earthquake"—firefighters, police, and rescue workers. So he was moved from the third game to the fourth, Harold and I from the fourth to the fifth. I had seen the first two games of the series, which the Giants had lost to Oakland in a woeful manner. Fearing that the Giants were going to be eliminated in the minimum four games, I protested the change to the then commissioner of baseball, Fay Vincent, and asked instead that Harold and I be allowed to share the mound with Willie Mays at game four. There was a short pause and then the commissioner said, "Doc, get real."

The Giants fell in four. Harold and I had to settle for a consolation prize: the first pitch at a Giants-Dodgers game the following season. I had been living in hope of (but not practicing for) that moment since the age of ten. In front of 38,000 people, I unloaded a one-hopper to the catcher, Terry Kennedy, who then ran out from behind the plate to shout: "You should let go of the ball earlier, Doc." True to form, Harold threw a perfect strike. I later learned that he had practiced on a regulation pitching mound. The Giants won, 4–3. I still have the ball.

I arrived back at my seat in the stands, to be greeted by the good-natured razzing of the fans who now knew me for what I was. I apologized to my two sons, who had approached the spectacle with great expectation. Elder son Dylan came through again: "Come on, Dad; the only thing that matters is that you were out there." Which brings to mind an aphorism attributed to Woody Allen, to the effect that "95 percent of success in life comes from just showing up." Nevertheless, I find myself living with enlarged expectations without enlarged talents. Students expect more of me, my colleagues expect more of me, I expect more of myself. But it is not to be had: I am what I have always been.

The Nobel Prize has not relieved me of the gnawing suspicion that prizes for science, prizes for any creative endeavor, have little merit. Here I have notable company, including the poet Ezra Pound ("It is extremely important that poetry be written, but it is a matter of indifference who writes it"), and the biologist and Nobel laureate Peter Medawar ("Scientists are always dispensable, for in the long run, others will do what they have been unable to do themselves"). The scientist and Pulitzer Prize–winning author Jared Diamond has argued that the last two centuries have produced only two scientists who might be deemed "irreplaceable"—Charles Darwin and Sigmund Freud.[41] Diamond admitted a temptation to also include Albert Einstein, but failed to explain why he did not. His choice of Freud will not sit well with many biological scientists, but I have no quarrel with it.

I have no illusions. I know that my good fortune could easily have gone to someone else. The title of this book was meant to be facetious. I have no formula for winning the Nobel Prize. I gave the prospect little thought until Dylan came into my bedroom early on October 9, 1989.

There were others who also thought little of my prospects. Soon after receiving the Nobel Prize, I was shown a written wager in which a distinguished professor at the University of California had once given three to one odds against my ever becoming a Nobel laureate. The wager also allowed for the unpleasant possibility that I might not live long enough for the issue to be resolved, which would have left Harold to test the handicap (or advantage) of death for Nobel honors. The bet was paid, at a considerable sum. I later managed to rub salt in the wound by projecting a photograph of the original document during a lecture attended by the loser. My effort at humor was not well received, my need for further lessons in discretion confirmed.

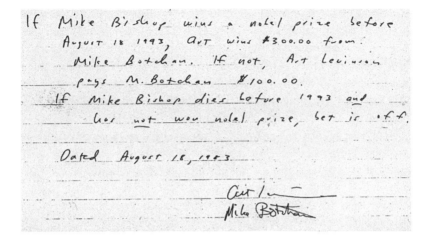

Nobel wager. The text of a wager on the author's chances for a Nobel Prize. The wager was made public during a lecture on October 17, 1989, twelve days after the announcement that the author and Harold E. Varmus would receive the Nobel Prize in Physiology or Medicine. Few remember the revelation because it was made during the hour that the Loma Prieta earthquake struck San Francisco.

The title of this book notwithstanding, I have an aversion to using the word "win" when speaking of the Nobel Prize. The verb inspires a competitive view of science that I find repugnant. Harold Varmus and I ran no race. We did our work and it happened to lead to an amazing place. There was no talk of the Nobel Prize when we began our experiments and none when we had our discovery in hand—the full significance of the discovery was not immediately apparent.

James D. Watson has told a different story in his widely read account of life in science, *The Double Helix*.[42] Watson portrays himself as preoccupied with pursuit of the Nobel Prize during his historical work on DNA with Francis Crick. But Crick has since said that he does not remember things that way: "If [Watson] really was thinking about Stockholm he must have kept it strictly to himself . . . [He] appeared strongly motivated by the scientific importance of the problem . . . It didn't occur to me that our discovery was prizeworthy until as late as 1956 [three years after the first publication] and then only because of a casual remark [made by a colleague]."[43]

However it was for Watson and Crick, Harold Varmus and I never discussed our prospects until after they had been realized, and like Crick, I too had not given the matter any thought until others started mentioning it to me. The first such mention is permanently engraved in my memory. I was sharing a small dormitory room with a younger scientist who had once worked with Harold Varmus and me. I was lying in the upper level of a bunk bed, my roommate in the lower level. We were both on the brink of sleep when he spontaneously announced: "You realize, Bishop, that you will be going to Stockholm"— or something to that effect. I was stunned by his certainty.

Scientists sometimes do find themselves in brisk competitions, either for discovery or for credit after the fact. But such competitions were occurring long before prize-giving had invaded the world of science, and they would continue to occur even if the Nobel Prize and its ilk were banished from the earth (a measure once proposed by an editorial in the journal *Nature* and received with notable silence).[44] Winning prizes is not the point of science; it is not the objective of most scientists. We do our work because we are enthralled and challenged by the puzzles of nature, because we can think of nothing else that we would rather do (although I might well now choose to play with a fine string quartet, if I had the requisite talent; or to pitch for the Giants, for which I clearly do not have the talent). My revered peers are not still congratulating me; they are asking me: "What is new?" If my answer is not satisfactory, my feet seem ever more like clay. Like the literary critic Van Wyck Brooks, "I feel every morning that I am on trial for my life and will not be acquitted."[45]

Scientists obey a demanding ethos, articulated by the poet William Butler Yeats:

> The intellect of man is forced to choose
> Perfection of the life, or of the work,
> And if it take the second must refuse
> A heavenly mansion, raging in the dark.
> When all that story's finished, what's the news?
> In luck or out the toil has left its mark:

That old perplexity an empty purse,
Or the day's vanity, the night's remorse.[46]

The scientist almost inevitably chooses perfection of the work, and in this age, that does not necessarily threaten "an empty purse." But some do hedge their bets. I am reminded of Gordon Tomkins, a distinguished colleague who died tragically and prematurely some years ago. Gordon was both an imaginative scientist and a talented musician. He was fond of telling new acquaintances of how he delayed his choice between science and music until after the age of thirty (here followed a pregnant pause), and was still wondering what might have happened if he had chosen science. In a more modest way, I have lived my life in the same straits. That story follows.

CHAPTER 2

Accidental Scientist

A [person's] first duty, a young person's at any rate, is to be ambitious, [and] the noblest ambition is that of leaving behind one something of permanent value.

—*G. H. Hardy*, A Mathematician's Apology

Danse de Biochemiste by Lars Bo, ca. 1964. (Reproduced by permission of the estate of the artist.)

I reached my own life in science by a twisting trail.[1] I like music as much as science—perhaps more. While I was still young, however, I realized that I had more talent for scholarship than for music. So I took what I had and ran with it. But it remains a small miracle that I ever became a scientist at all. And it remains a matter of some regret for me that I could not become a musician.

I was born in 1936 and spent the first fourteen years of my life in a town of four hundred in rural Pennsylvania. Those years were pastoral in two senses of the word: I saw little of urban life until I was past the age of twenty-one, and my youth was pervaded with the concerns of my father, a Lutheran minister tending two small parishes. My pastoral years endowed me with two durable legacies: a sense of wonder that began with a youthful attachment to biblical tales but that transmuted easily into wonder at the natural world; and a passion for music, sired by the liturgy of the church and fostered by my parents through piano, organ, and vocal lessons. I am deeply grateful for these legacies, albeit apostate from the church.

Early Education

I received the first eight years of my education in a two-room school. There I came under the influence of a remarkable man who taught all of the subjects in grades five through eight in a single room. He was a rough-spoken, stern, but compassionate person whose contract called for a school holiday on the first day of every deer hunting season. If he failed to get a deer on the first day, the contract extended the holiday one additional day. His students cheered for the deer on the first day, but for the teacher on the second—he could be fearsome when frustrated.

The deer hunter was an engaging teacher who awakened my intellect with instruction that today would seem rigorous in many colleges. History figured large in the curriculum, exciting for me what would become an enduring interest. He was a fierce disciplinarian and strict grammarian, occasionally taught calculus to eighth graders, and enforced penmanship with a vengeance that preserved my handwriting through even the worst hazards of a medical education. But I can recall no instruction in science of any sort during those first eight years of schooling, other than the collection and pressing of wildflowers.

My high school was also small. Only a few of my sixty or so classmates went on to college, and only I became a scientist. None of my high school faculty were deer hunters. But they all offered encouragement as my intellectual capabilities continued to emerge. At my graduation ceremony, I made a point of thanking the track coach for my precious varsity letter (which I have kept to this day). He countered that my place as valedictorian of the class had far greater significance, and that he would rather be thanked for the physics course he had taught me. I can remember his vivid rendition of the Michelson-Morley experiment to this day.[2]

Shortly after I received the Nobel Prize, I lectured at a medical school not far from my home town. Following the lecture, a figure emerged from the audience whom I eventually recognized as that track coach and physics teacher from my youth. He had ridden his motorcycle more than fifty miles to hear me speak, even though he knew that the subject would be beyond him. I had not laid eyes on him since that evening of graduation, yet we struck an immediate rapport that had eluded us when I was an adolescent and a woeful—albeit determined—athlete. Although I was a member of the mile relay team that won the Pennsylvania state championship in its class, I ran in second position, tactically reserved for the slowest member of the team. But now I had fulfilled his valedictory admonition to me and he was clearly pleased.

I had taken easily to school and was an excellent student from the beginning. My only difficulty came in the second grade, when I found myself bored with the pace and, thus, was even more hyperkinetic than usual. Weary of my interruptions, the teacher summarily assigned me to the third grade, and that kept me out of trouble. Nowadays, medication might well have substituted for promotion.

Paradoxically, my aspirations for the future were formed outside the classroom, when I was befriended by Robert Kough, a physician who cared for members of my family. Although practicing general medicine in a rural community, he possessed remarkable intellectual vigor and rigor. He aroused in me an interest not so much in the life of a physician as in the fundamentals of human biology. I recall picking up a medical journal in his office one day and finding a lead article on the management of pilonidal cysts. Even at the age of thirteen, I sensed that this would not be for me.

College

Still, I entered Gettysburg College with medical school in mind. But my ambition was far from resolute. Every new subject that I encountered in college proved a siren song. I imagined myself a historian, a philosopher, a novelist, occasionally a physician, but never a scientist (in part because I then had no idea of what a scientist might do). And my horizons were expanding. On learning that neither I nor several of my close friends had ever been inside an art museum, our professor of German promptly hauled us off to the National Gallery in Washington, D.C., a three-hour drive, for a crash course in how to look at paintings. The tour ended in front of the renowned (and often reviled) *Last Supper* by Salvador Dalí. The venom of the critique delivered by our professor was memorable because I had never before encountered such passion over aesthetics. I have been looking at works of art with joy (and occasional disapproval) ever since.

Despite my intellectual wanderlust, I stayed the course, completing my major in chemistry with diffidence and the bare minimum of credits. I met the woman who was to become my wife, Kathryn Ione Putman. (She is still with me more than four decades later, a matter of some pride in California.) I have never been happier before or since.

But I still saw nothing of research. Gettysburg was a small liberal arts college that valued creativity, but in those days provided no opportunities for laboratory research, nor did it occur to me at the time that it should. Nevertheless, it was one of my college faculty who first challenged the strength of my vocation. During an informal discussion, my revered professor of physics abruptly interjected: "Why do you want to be a doctor? They are nothing but well-trained plumbers." That hyperbolic challenge shook me to the core.

Years later, I told this story to an interviewer from the college's alumni magazine. They promptly published it, infuriating every physician who read the magazine. My professor never forgot my imprudence, although until his death a few years ago, he continued to receive me with affection whenever I returned to my alma mater.

As I look back on it now, one point above all others moves me. To touch a life the way so many of my teachers touched mine is a privilege, a responsibility, an opportunity not to be dismissed lightly. It makes a singular case for teaching as a gratifying and vital career.

Medical School

Eventually, I had to choose a medical school. At the appointed time, my chemistry advisor called me in to ask what I might want to do with a medical education. I answered that I had become interested in the life of an academic,[3] but still intended to attend medical school because nothing else had caught my fancy quite strongly enough.[4] "Then you should consider Harvard," he advised. "Where is that?" I asked. "In Boston somewhere, I think," was his response.

I eventually found the correct address and applied, adding the University of Pennsylvania for good measure. Making only two applications was probably foolhardy even in those days; in the present, it would be self-destructive. Both schools admitted me, the University of Pennsylvania by means of a letter sent regular mail, Harvard by means of a telegram. The contrast foreshadowed what was to come. (Truth be told, I also applied to Johns Hopkins University School of Medicine, but withdrew my application after visiting the school—my rural sensitivities were not yet prepared for the gritty realities of the Baltimore neighborhood in which the school resides.)

I prepared my applications while working as a summer employee in

Unfocused. The author in Wyoming at the time of his application to medical school. (From the family album of the author.)

Yellowstone National Park. I wrote by hand with a ballpoint pen, and included a statement that I sought a career in medicine because I thought this would provide me both gratification and a comfortable living. No contemporary medical school of any note would now entertain an application unless it were flawlessly typed (and preferably, electronically submitted), and few would look kindly upon a crass confession of material aims—applicants are expected to endorse abnegation and fierce social purpose.

How to choose? My interview for Harvard had been in Philadelphia, so I still knew virtually nothing about the school other than that it was for some mysterious reason celebrated. I wrote a letter to Harvard, explaining that I was having difficulty deciding between it and the University of Pennsylvania. Could I come and visit? Years later, the dean of students at Harvard told me that my letter had been posted in the dean's office for the amusement of the staff. Thus did I learn the measure of institutional arrogance.

The visit to Harvard came to pass and was little short of a fraternity rush (a memory that flatters me even now, in the face of all my cynicism about institutional arrogance). As host, Harvard assigned me a savvy Ivy League tennis champion who discussed Schopenhauer with me at breakfast. Neither he nor I actually spoke much sense in that conversation, but we both had a good time. He took me to see open-heart surgery. He briefed me on tweed jackets. I was beginning to get the point.

Then I got some unexpected assistance from an associate dean at the University of Pennsylvania, who was interviewing me for a scholarship. On learning of my academic aspirations, he recommended that I decline my admission to his school and attend Harvard. I have rarely encountered such candor since. The point had now been fully made.

So in the autumn of 1957, I went off to Harvard Medical School, which was indeed in Boston. That city proved to be a revelation and a revel, replete as it was with music and fine arts. Harvard, on the other hand, was a revelation and a trial. I discovered that the path to an academic career in the biomedical sciences lay through research, not through teaching, and that I was among the least prepared among my peers at Harvard to travel that path.

The people who mattered most to me as I grappled with this revelation were several sophisticated classmates whose prior experience allowed them to teach me the ethos of research. They became my principal mentors throughout medical school and enduring friends. So it is that, to this day, I tell my students that they can expect to learn more from their peers than from their faculty.

The ethos at Harvard catered to intellectualism and further discouraged me from any inclination toward the practice of medicine. Research was portrayed as the most esteemed of medical endeavors, a state of grace to which all should aspire (much to the annoyance of many of my classmates, who understandably had thought that medical school was mainly about becoming a doctor). So I sought out research experience in a neurobiology laboratory, but was rebuffed because of my inexperience. I became ambivalent about continuing in medical school, yet at a loss for an alternative.

Finding Research

During my second year in medical school, two pathologists rescued me from my dilemma. Benjamin Castleman offered me a year of independent study in his department at Massachusetts General Hospital, and Edgar Taft of that department took me into his research laboratory. There was little hope that I could do any substantive investigations that year, and I did not. But I became a practiced pathologist, which gave me an immense academic advantage in the ensuing years of medical school. I found the leisure to marry. And I was riotously free to read and think, which led me to a new passion: molecular biology, which was then just beginning its triumphant sweep through medical science. I have never had such autonomy before or since, and I credit the autonomy for making that year the most important in my life (there was also my marriage, of course).

I began to teach myself what I might need to know to become a scientist. I did this mainly by making regular visits to the premier medical bookstore in Boston and bringing home haphazard assortments of books, which I read according to whim. Kathryn and I were living on her slender income as a public school teacher because Harvard had cancelled my scholarship when it learned of our marriage—spousal income, however scant, was regarded as a due substitute for Harvard's benefaction; the university thought it sufficient that I retain the title of National Medical Scholar, without the stipend. But once within the confines of that bookstore, I became oblivious to budget. I had initiated a mania for books that has never slackened, and a selective disregard for frugality that has served my mania well. I still have all of the books acquired in that year. Few are now worth the paper on which they are printed, but they stand on my shelves as mementos of a turning point in my life.

Self-instruction follows an honorable tradition, even when as undisciplined as my own. I once heard Freeman Dyson remark that he

Laboratory Still Life No. 4 by Tony Cragg, 1988. (Reproduced by permission of Crown Point Press.)

had learned much more about science as a child from reading books and visiting museums, than from formal instruction.[5] Granted, this might not work for others: Dyson has never for a moment been a mere mortal. Indeed, looking back over my own career and its failures, I cannot help but wonder whether I have suffered unduly from being an autodidact in almost everything that I tried to master, from research to fly fishing. Might formal training have made me better? I believe I know the answer, and it is disquieting.

Whatever its limitations, my year of autonomy set my course toward research. And I was gradually becoming shrewd. I recognized that molecular biology had advanced far beyond my existing capabilities, that its inner sanctum was not accessible to one so unsophisticated as myself, that I would have to find an outer chamber in which to pursue my passion. I found animal viruses, those tiniest of creatures that can wreak such havoc with human health—the annoyance of the common cold, the global mortality of influenza, the horror of smallpox, the modern plague of AIDS.

Animal viruses came to my attention through an elective course taken when I returned to my third year of medical school. Elmer Pfefferkorn, at the time an unsung instructor who taught the course, took me into his miniscule laboratory and put me to work. Elmer soon rose to great distinction in his field and eventually became chairman of the Department of Microbiology at the Dartmouth School of Medicine. I readily concede that my work with Elmer contributed nothing to either of those achievements. From the course, I learned that animal viruses were ripe for study with the tools of molecular biology, yet still accessible to the likes of me. From Elmer, I learned the exhilaration of research, the practice of rigor, and the art of disappointment.

I began my work with Elmer in odd hours snatched from the days and nights of my formal curriculum. But an enlightened dean of students gave me a larger opportunity when he approved my outrageous proposal to ignore the curriculum of my final year in medical school so that I could spend most of my time in the research laboratory. The only requirement was that I explain myself to the chairs of the various departments whose offerings I would be ignoring. That made for some interesting interviews. But no one blocked my way. (I realize now that the dean had, to a modest extent, passed the buck. But no matter: it worked for me.)

In the end, I completed only one of the eight or so formal courses then required of fourth-year students. Flexibility of this sort in the affairs of a medical school is rare, even now, in this allegedly more liberal age. In most states (California included), it would be a statutory impossibility because of legislative requirements that constrain the medical curriculum in ways that defy reason and wisdom. It has been more than thirty years since Christopher Jencks and David Riesman concluded that "there may be almost no causal relationship between learning what is taught in professional school and doing well as a professional practitioner."[6] This is an insight that I can affirm, but that medical education in the main continues to ignore.

My work with Elmer was sheer joy, but it produced nothing of substance. I chose not to submit a thesis based on my unsuccessful experiments, a decision that was later cited to me as the reason I had graduated cum laude rather than magna cum laude (an obvious wound, else I would not still remember it). I remained uncredentialed for further work in research. So almost by default, I entered an interregnum of two years as a house physician at Massachusetts General Hospital. That magnificent hospital admitted me to its prestigious training despite my woeful inexperience at the bedside, and despite my admission to the chief of the medical service that I had no intention of ever practicing medicine. I have no evidence that they ever regretted their decision. Indeed, years later, I was privileged to receive their Warren Triennial Prize, one of my most treasured recognitions. I cherish the memories of my time there: I learned much about medicine, society, and myself; and I had put the lie to the confident predictions by peers, faculty, and departmental chairs that my wanton fourth year in medical school would doom my career.

Finding a Place

But I was aching for a return to the laboratory. On my final day as a medical house officer, as I walked out of the emergency ward toward a different sort of future, I removed the bulky pager from my belt and, in a moment of reckless euphoria, hurled it against the wall—disabling it beyond repair, I am sure. No one ever sent me a bill.

Clinical training behind me, I began research in earnest as a postdoctoral fellow in the Research Associate Training Program at the National Institutes of Health (NIH), in Bethesda, Maryland, which was designed to train mere physicians like myself in fundamental research. At the time, the program was a unique resource, providing U.S. medical schools with many of their most accomplished faculty. Without this assist, it is unlikely that I could have found my way into the community of science. Certainly, no academic laboratory brimming with well-trained Ph.D.'s was about to take me in. Times have changed. The

United States now abounds in programs that welcome newly minted physicians for research training, striving to reverse a steep and unwelcome decline over recent years in the number of physicians who pursue research.

I barely escaped the clutches of the U.S. Army, which put me through a full induction exam and was poised to draft me when my commission in the Public Health Service for work at the NIH finally arrived. So I joined the cadre of medical scientists who were sequestered from violence by their positions at the NIH under the disparaging sobriquet of "yellow berets." I never exercised my entitlement to wear the uniform of a lieutenant commander (or to travel gratis in military aircraft).

My mentor at the NIH was Leon Levintow, who has continued as friend and alter ego to this day—a relationship cemented as much by a common love of music and gossip as by shared interests in science. (Gossip is a common coin in scientific discourse, as explained by Francis Crick: "What you are really interested in is what you gossip about."[7]) Leon helped me in many ways. But preeminent among these was by being my advocate with administrators and scientists alike. He developed a confidence in my prospects and he made that confidence known in many useful ways, while I was at the NIH and in the years to come. Every young scientist can profit from such an advocate, and every senior scientist should be willing to be one. There is a remnant of Renaissance patronage in the practice of modern science that is both admirable and effective.

I began research on the means by which the poliovirus might reproduce itself. My friends outside of science were puzzled that I should devote myself to such work. After all, it was already clear that the vaccines developed by Jonas Salk and Albert Sabin would eventually eliminate the fearful paralytic disease caused by this virus (and by now have largely done so). So what was left to do? The answer lies in the role of simplification in the practice of science.

Despite its diminutive size, the mammalian cell is fiendishly complex, its lifestyle sustained by tens of thousands of genes and equally abundant chemical reactions. In contrast, most viruses have relatively few genes of their own (typically no more than a dozen), yet reproduce

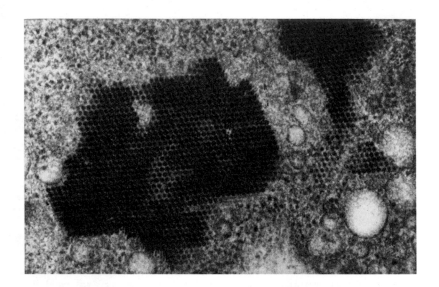

Poliovirus. Crystals composed of numerous virus particles within a human cell. Each black hexagon is an individual virus particle. Magnification approximately 20,000-fold. (Electron micrograph from collection of the author.)

in a bountiful way by parasitizing the machinery of a mammalian cell. When we explore the reproduction of those few viral genes, we obtain a sketch in miniature of the machinery that sustains the cell itself.

The strategy of simplification sired the molecular revolution that has transformed our understanding of life and death over the past several decades. Yet we still struggle to understand the molecular underpinnings of how poliovirus reproduces itself and causes disease. There is a paradox here. We often hit upon remedies for practical problems (such as the vaccines for poliovirus) before we achieve a fundamental understanding of the processes that underlie those problems (such as the ability of poliovirus to replicate and induce disease). But once the fundamental understanding is in hand, even better remedies can follow.

Working on poliovirus brought me my first publishable research. My feet were now thoroughly wet. I had found a place for myself. Or had I? People began to ask where I was going next. What future could

I imagine? I had no idea about this, had really never given it any
thought. Then departments of microbiology began to offer me jobs.
So I arbitrarily christened myself a microbiologist and soon realized a
natural attachment to the discipline. After all, it was the study of microbes that had spawned the molecular revolution in modern biology,
and that revolution had first lured me into research. To this day, however, I hesitate whenever asked to name my discipline. I am in fact a
dilettante. I could be nothing other. I enjoy every minute of it. But it is
in most eyes a disreputable fate.

Midway through my postdoctoral training, Leon Levintow departed
for the faculty at the University of California, San Francisco (UCSF).
In his stead came Gebhard Koch, a visitor from Germany with whom I
began to collaborate and who, in 1967, lured me to his home base in
Hamburg for a year. Once again, I had an enlightened benefactor: the
distinguished virologist Karl Habel, who appointed me a permanent
member of the NIH staff and then agreed to have the federal government provide my salary in Germany during the very first year of my
appointment. I repaid the benefaction by never returning to Bethesda.
I have been paying my taxes willingly ever since and whispering words
of gratitude to the body politic (as should every scientist who enjoys
research support from public funds).

During my year in Germany I had little success in the laboratory,
but I learned the joys of Romanesque architecture and German ex-
pressionism. As the year drew to a close, I had in hand my permanent appointment at the NIH, as well as two offers of faculty positions—one at a prestigious university on the East Coast of the United
States, the other from Levintow and his departmental chairman, Ernest Jawetz, at UCSF. Seizing the chance to realize my youthful ambition to be an academician, I abandoned the NIH and chose UCSF.

In those days, UCSF was hardly known outside the city limits of San
Francisco. When I told a friend at the NIH of my plans, he claimed not
to know that there was a medical school in San Francisco (let alone a
full health science campus, which was in fact the case). Yet my decision
to go there was an easy one, because the opportunities involved in going seemed so much greater than those in staying. I would have been a
mere embellishment on the East Coast. I was genuinely needed in San

Francisco. And not incidentally, I had fallen in love with the city itself. I have a vivid recollection of riding a bus during my first visit there. By the time I got off, I had heard six languages. I knew then that this was a place where I would want to live.

In February of 1968, my wife and I moved from Hamburg to San Francisco, where we remain ensconced to this day. That decision was so successful that it still underpins my advice to young scientists who are in the market for a job: go where you are genuinely needed; do not let prestige set the course.

The Academician

Now I faced that rite of passage for all young university scientists: the first application for a research grant. At the time, the NIH was in one of its periodic nadirs of funding. So the elders who read my proposal in advance of submission cautioned that it was far too ambitious, that I had asked for too much money, and that there was no way on God's earth that I would get the five years of funding that I had requested. There is no doubt that my proposal was wildly, unrealistically ambitious. But I got the grant, every penny that I asked for, and for five years. A budget officer at the NIH did try to talk me out of a pricey gadget known as a scintillation counter, but I stood my ground and prevailed. The moral from all this is that there is no harm in asking.

I continued my work on poliovirus. But new departures were in the offing. In the laboratory adjoining mine, I found Warren Levinson studying Rous sarcoma virus, which causes cancer in chickens and rodents.[8] Rous sarcoma virus is an archetype for a large family of viruses now known as "retroviruses." (A latter-day representative of these viruses is HIV, the cause of AIDS). At the time, the mechanism by which retroviruses reproduce themselves was one of the great puzzles of medical research. Levinson, Levintow, and I joined forces in the hope of solving that puzzle. I soon realized that something very strange was afoot.

The work of Howard Temin at the University of Wisconsin had raised the possibility that retroviruses might do the unthinkable. They seemed to reverse the flow of genetic information, allowing it to go backward through what was universally thought to be an inviolable and unidirectional molecular circuitry (see Chapter 4 for a description of this circuitry). This was heresy, for which Temin was widely ridiculed. But while preparing lectures for graduate students during my first year in San Francisco, I reached the conclusion that Temin was probably right, and that the molecular machinery to implement his heresy must reside inside the virus itself.

In early 1969, I pursued the thought with several experiments but abandoned the chase prematurely, out of concern that I was not pursuing the stated objectives of my grant from the NIH, and in response to skepticism on the part of several of my older, more experienced colleagues. I know now that my concern was misplaced: most research grants are hunting licenses to bag anything of value—one of the great strengths of the U.S. research enterprise. Too many young scientists are still advised otherwise by their elders. As for skepticism by those elders, it should be taken as a goad to persevere.

As it turned out, I had stopped just short of success. I learned this one year later, when Temin and David Baltimore at the Massachusetts Institute for Technology announced that they had independently discovered in retroviruses an unprecedented enzyme that reverses the usual direction of flow for genetic information and allows the retrovirus to parasitize cells in perpetuity.[9] The enzyme was quickly dubbed "reverse transcriptase." It was a momentous discovery that entered textbooks immediately. Neither Temin nor Baltimore had known what the other was up to until both had announced their discovery.

The Nobel Prize followed for Temin and Baltimore a mere five years later. And I acquired new respect for the hand of fate. During his years in high school, David Baltimore had attended a summer camp for promising science students. His counselor at that camp was Howard Temin, then a student at Swarthmore College. Baltimore followed suit by also attending Swarthmore. The two then went their separate ways until a day years later when they spoke on the phone to confirm that each had authenticated the same heresy.

The discovery of reverse transcriptase was a devastating blow to me. A momentous secret of nature, mine for the taking, had eluded me. I grieved for months; I still grieve in weaker moments. I had learned three lessons the hard way. First, the outsider often sees things more clearly than the insider and should not be intimidated by his inexperience. Second, the scientist must trust her or his own imagination, even if, perhaps especially if, it runs counter to received wisdom. Third, there is no substitute for intellectual daring: if you want to rise above the pedestrian, you must be prepared to take risks.

These lessons differ very little from the elements of artistic creativity, not commonly thought to resemble scientific talent. In many ways, however, art and science are kindred souls. Both arise from the same transcendent human qualities: ambition, imagination, creativity, intellectual daring, and the urge to discover. The critic Roger Lipsey once described the forces that drive an artist: "Knowledge permits the artist to work, conferring confidence and direction. But ignorance joined with longing and curiosity draws the artist forward, motivates, authorizes free experimentation and play."[10] It would be difficult to write a better recipe for discovery in science.

It is telling that the very word "scientist" was coined by analogy with "artist." The coinage came in an anonymous book review by William Whewell in 1834. He devised the word for want of a term that encompassed all the sciences, and he credited the physical scientist Mary Somerwell, a "[person] of real science," with the inspiration.[11] Still, I find myself living in some envy of the artist. We scientists are slaves to the puzzles preformed by nature and to our rules, whereas artists create their own puzzles and solve them by breaking rules.[12]

So I mourned my failure. But I was also exhilarated because reverse transcriptase offered new approaches to the study of retroviruses, approaches that I seized and deployed with a vengeance. I was joined in this work by a growing force of talented postdoctoral fellows and graduate students.[13] Preeminent among these was Harold Eliot Varmus, who arrived in my laboratory as a postdoctoral fellow in late 1970. This too was inadvertent. Harold had not chosen me nor I him. He had been deflected to me by a senior figure in the field who apparently thought that neither Harold nor I deserved any better.

Harold's arrival changed my life and career. Our relationship evolved rapidly to one of equals, and the result was surely greater than the sum of the two parts—unless I have underestimated Harold. For nearly fifteen years, we jointly supervised a group of younger scientists that at its peak numbered more than two dozen. The two of us met regularly with each of them, exchanging ideas, reviewing data, criticizing conclusions, and helping to write the manuscripts that would convey our news to the scientific world. We were, in essence, two organists playing on the keys of a splendid instrument. Except that these keys had spirited minds of their own, so dissonance was not unheard of, and on many occasions, the keys actually drafted the score and then prevailed over the organists. The arrangement between Harold and myself was unusual, widely recognized as such, and much admired. We became a hyphenated self that gave its name to a social organism— the "Bishop-Varmus" laboratory. Among my generation of biomedical scientists, I know of very few such partnerships that achieved comparable distinction.

I am often asked what cemented our relationship through the years that we were a team. The immediate adhesive was a shared infatuation with science in general and cancer viruses in particular. But the bond was set more strongly by our mutual love of words and language. We are both voracious readers and enjoy writing. In contrast, we are very different as scientists: Harold revels in detail; I am impatient with it (and that has cost me dearly).

The partnership between Harold and myself was still young when lightning struck. We unexpectedly discovered a way to pry open the black box that had hidden the inner secret of the cancer cell. Through our study of Rous sarcoma virus, we were led to a group of normal cellular genes whose malfunction can produce cancer. Many causes of cancer may all wreak their mayhem by damaging these genes, converting them to cancer genes. One of my great moments of professional fulfillment came when I later encountered a description of our discovery for the first time in a college text. Soon thereafter, it began to appear in high school curricula as well, and I found myself tutoring my own sons about the concept. Harold and I had fulfilled the duty laid out by G. H. Hardy in the epigraph to this chapter.

Chapter 4 will tell the story of our discovery in some detail.[14] Suffice it to say here that without design or warning, I had become a cancer researcher. The ensuing years have been consumed by the pursuit of how cancer genes function and how they can contribute to the genesis of cancer. My great hope was that I could bring my own research directly to bear on human cancer. My failure to do this in any consistent way has been my greatest disappointment.

What have been the satisfactions of my unexpected life in science? Pursuit of discovery was certainly the headiest in the early years of my career. But I confess that over time this has dwindled, with the inevitable and painfully clear realization that lightning rarely strikes twice in the same place. There have been other compensations, however, particularly the pleasure of helping others reach their potential. Working with a group of younger scientists toward a shared goal is immensely gratifying, even when it brings only small successes. Every knock on my office door that heralds an unexpected piece of data is a reaffirmation of life.

And always, there has been teaching, an integral part of the academic career no matter what the setting—a mandatory part, an essential part, an honorable part, a gratifying part of the career. Why might scientists feel the obligation to teach? To answer that question, I recall the debt I owe to the teachers I have known at every step in my education. They helped shape the aspirations, talent, and discipline that have brought me lasting gratification in my profession. But more important, they helped me see that the intellect is among our most distinctive and precious possessions; and the exercise of intellect, one of our greatest pleasures. Who would not want to do that for others?

The desire to teach is visceral: it requires no defense, it permits no explanation, it is a cultural obligation, it is a vocation. Scholarship and research without the vocation to teach are sterile. There is another view on this matter, however, once articulated by John Henry Cardinal Newman: "To discover and teach are distinct functions; they are also distinct gifts, and are not commonly found united in the same person. He who spends his day dispensing his existing knowledge to all comers is unlikely to have either leisure or energy to acquire new . . . The greatest thinkers have been men of absent minds and idiosyncratic habits, and have more or less shunned the lecture room and public schools."[15]

What a malign view of our estate! There is nothing more wonderful in all of human experience than the "absent minds" and "idiosyncratic habits" of great thinkers. These glories should be on display in the classroom. As for those of us with lesser minds, if we are held in thrall by the unending surprises of nature and if we are locked in pursuit of those surprises, we can bring to teaching a fire few students can resist. The noblest calling of a modern academic is to combine the distinct crafts of discovery and teaching in the same person.

Lessons

My life in science has taught me to distrust premeditation. Recall that I entered college without a clear vocation. Those infamous "preference tests" of the time predicted that I might become a journalist, a musician, or a forest ranger. There was some reality in those predictions: I have a special affinity for newsprint, I begrudge every day that passes without the sound of live music, and I have a largely thwarted but still passionate attachment to wilderness. But none of these prefigured my professional destiny. Similarly, I began medical school with little interest in practicing medicine. Slowly but surely, under the influence of fellow students and a few memorable teachers, I found my way to research, launched on a career only after passing the age of thirty.

Harold Varmus proved an equal if not greater exception.[16] Like me, he entered college intending to study medicine. But he gravitated to English literature and journalism, and failed to distinguish himself in science. At one point, he was advised to leave a class in organic chemistry because of his poor performance. The advice was ignored. (It may be just as well that the advisor did not live to see Harold become both a Nobel laureate and, for six years, director of the NIH.) Harold then obtained a master's degree in English literature at Harvard before finally opting to attend medical school.

But nothing was coming easily. Harvard Medical School rejected Harold twice, with the advice from one interviewer that he join the army in order to acquire some "focus." The College of Physicians and Surgeons at Columbia University displayed greater prescience by accepting Harold and setting him on his future course. He too established himself in research only after the age of thirty. His career thereafter makes the case for being unfocused in one's twenties (as does my own, I suppose).

The paths that Harold and I have followed were not premeditated journeys to a calculated goal. We followed our noses and they led to an amazing place. The lesson here for me is that those of us who teach should place less emphasis on recitation and more on inspiration. We should educate and influence, but we should also let our students follow their own noses.

Along the way, I learned how important it can be to have great personal resolve, to cultivate colleagues, to ignore convention, and to look for new vistas. The last of these—the search for new vistas—should hold a special place in the lives of young scientists. I was privileged to participate in the birth and maturation of two research fields, and in both, the great exaltations came mainly in the beginning. It is the pioneers in science who have the most fun (albeit not always the most fame).

There is a special trick to the selection of new vistas that Jonathan Weiner has called "Occam's Castle" (by analogy with Occam's Razor— the strategy of choosing the simplest available solution): "Faced with several competing places to build a new science, prefer the simplest one. Pick the place that requires the least preparation, the least digging, hauling out, pouring in, and shoring up."[17] In other words, start something truly new: "Things are always best in their beginning."[18] The distinguished biologist Sydney Brenner concurs: "It's what I enjoy most, the opening game. And I'm afraid that once it gets past that point I get rather bored and want to do other things."[19] Boredom can be a tonic for the scientist—a sign that it is time to move on.[20]

Above all else, however, I have learned that there is no single path to creativity, not even within the stern halls of science. We are constrained not by the necessary discipline of rigor, but by the limits to our imaginations and to our intellectual daring. The artist Ben Shahn said it most succinctly: "Form is an instrument, not a tyrant."[21]

Frenzied activity has become a fetish of modern scientists, many of whom speak proudly of sixteen-hour days, schedules honed to the minute, and more travel than that of Alfred Nobel himself. But frenzy is the enemy of reflection, and reflection is central to discovery. Time

and tranquility permit the intellectual synthesis and leaps of imagination that generate insight. Francis Crick has warned against the temptation to "work so hard that there is no time left for serious thinking
. . . [Scientists] should heed the saying 'A busy life is a wasted life.'"[22] I
have had my own struggles with these realities. There is something in
my nature that compels me to regard each spare moment as a waste,
even knowing that the moment should be an opportunity to muse.

None of us can be perpetually inspired. I have an acquaintance who
many years ago asked a Nobel laureate of great renown to serve as his
thesis advisor. Visibly morose, the laureate declined. When asked why,

"Leonardo's Lament." © 2002 Sidney Harris. (Reproduced by
permission of the artist.)

he explained that he could "not think of anything to do." This is not a plight that Nobel laureates readily admit, nor one that students readily accept. My acquaintance went elsewhere, to a supervisor who was keeping dozens of young associates busy (and himself later became a Noble laureate who, in due course, also ran out of things to do).

Discovery takes two forms. The first is mundane, but nevertheless legitimate: we grope our way to reality and then recognize it for what it is. The second is legitimate, but also sublime: we imagine reality as it ought to be and then find the proof for our imaginings. I have been fortunate to know the first form of discovery and am thankful for the privilege. I have miscarried opportunities to know the second and am diminished by the failure.

Why Science Succeeds

There is a special truth about science that seems not to be widely appreciated. The success of science requires individual talent, but it is driven by personal values. Preeminent among these values is honesty. Scientists depend on the truthfulness of their colleagues. Each of us builds our discoveries on the work of others. If that work is false, our constructions fall like a house of cards and we must start all over again. Little wonder, then, that science places high value on the reproducibility of discoveries, whether they are dramatic or mundane.

Whatever value scientists might place on reproducibility, it is not shared by the editors of most research journals. They typically decline to publish work that they regard as only "confirmatory." The research community seems disinclined to resolve this inconsistency; indeed, seems to endorse it. Yet credible failure to reproduce a set of results is often publishable, particularly if it challenges a previous claim of some importance, and scientists rely on such publication to prevent them from investing good work after bad.

The success of science also depends heavily on equality. Science is at its best only when it welcomes talent of every description, treats every idea with equal respect, judges all ideas by a common set of standards. As individuals, scientists are often intolerant and unfair.[23] But in the rules of evidence that we apply to our practice of science, equality must prevail. When it does not, science easily goes astray—indeed, often embarrasses itself. And often the embarrassment arises when members of one discipline prove intolerant of ideas from another.

It was an astronomer and meteorologist, Alfred Wegener, who proposed that the continents arose from the fracture of a single global land mass (Pangaea) and have since been drifting on the surface of a molten earth. He was derided by an entire generation of geologists and geophysicists, but he was correct. The chemist James Lovelock first proposed that a self-regulating organic equilibrium sustains all of life on earth, causing our planet to simulate a gigantic organism (Gaia). Decades passed before biologists and earth scientists conceded some element of truth to Gaia. Howard Temin, a virologist, suffered years of derision before his claim that genetic information flowed in more than one chemical direction was vindicated and accepted by biochemists and molecular biologists. The moral from all this: "It is no accident that where the stranger is welcome, there is both tolerance and genius."[24]

Science thrives on the spirit of community. The popular mind imagines the scientist as a lonely genius. In reality, few of us are geniuses, and even fewer are lonely. Most scientists do virtually nothing alone: we exchange ideas with alacrity; we design and perform experiments together; we rely on one another day in and day out; we usually take pleasure in discoveries, no matter who has made them; we usually give credit where it is due. The popular press dramatizes our competitions. But for each of these, there are countless collaborations. Science spans all boundaries, creating what Freeman Dyson has called "a territory of freedom and friendship in the midst of tyranny and hatred."[25] Dyson wrote with great passion about friendship among scientists. "Scientists are as gregarious as termites. If the lives of scientists are on the whole joyful, it is because our friendships are deep and lasting. Our friendships are lasting because we are engaged in a collective enterprise."[26]

Ambition provides much of the energy required to practice science—ambition for personal achievement, for validation of personal worth through recognition, for "leaving behind one something of permanent value."[27] My upbringing left me with a puritanical suspicion of ambition as something ignoble and potentially corrosive, and so it can be. But ambition can also foster transcendant achievement.[28] There is really no other way to account for human advance than through ambition: it is the wellspring of creativity and diligence; it is the "first duty" of every young person.

Creative science is achieved through courage. Most of the great discoveries in science come from bold acts of the imagination, intellectual daring of the highest order. In the words of Fats Waller (speaking of musical chords): "Dare to be wrong, or you may never be right."[29] There is no fear in science greater than that of being wrong. But the scientist who cannot act in the face of that fear stands little chance of changing textbooks.

Aesthetics permeate science. Scientists find beauty in every nook and cranny of the natural world. It is their inspiration to work. The French mathematician and physicist Henri Poincaré once voiced this inspiration: "The scientist does not study nature because it is useful; he studies it because he delights in it, and he delights in it because it is beautiful. If nature were not beautiful, it would not be worth knowing, and if nature were not worth knowing, life would not be worth living."[30]

But the aesthetics of science are not always reliable. Scientists can be misled by the beauty of their theories or even of their supposed facts. It was once believed that all of the nerve cells in the brain were continuous with one another (they are not). The great histologist and Nobel laureate Santiago Ramon Y Cajal adhered to this view for many years because he found the theory beautiful. He eventually realized the error of his ways. "As always," he wrote in explanation of his error, "reason is silent before beauty."[31]

Political Scientists

Laboratory scientists have traditionally disdained politics. In doing so, they ignore two fundamental equations: to the extent that science relies on public support, as it clearly does, politics is essential to science; and to the extent that science supports the public welfare, as it clearly does, science is essential to politics. We ignore these equations at our peril. Science is no longer a thing apart: it has become part and parcel of our culture, a prevailing force of our time. So there is good reason for scientists to be mindful of politics, and for politicians to be mindful of science.

Several years ago, while attending an international meeting of scientists at the Moscone Conference Center in San Francisco, I was stationed in the front lobby to solicit an audience for a lecture by Senator Tom Harkin of Iowa—a distinguished public servant and a strong advocate of biomedical research. To my dismay, the response was not enthusiastic. In the words of one graduate student whom I happened to know: "I don't want to have anything to do with politics—it's dirty." Having delivered that zinger, she turned and walked away (perhaps oblivious to the source of the funds that were supporting her, her research, and even the conference that she was attending).

This stung, because I have another view based on an admiration for representative government and fortified by personal experience. My experience began in 1989, a time of declining funding for biomedical research by the NIH. Disillusionment, anger, even panic were widespread in the research community. One titan of industry assured me that the bulk of biomedical research would have to be privatized within a decade, that federal support would gradually disappear (it has in reality become stronger than at any previous time in our history).

Galvanized by these circumstances, a small group of biomedical scientists gathered in San Francisco to determine how we might become more active in the corridors of government. Our objective was to form a consortium of professional societies that would concentrate its energies on the funding of research grants by the federal government. Our constituency eventually grew to more than 25,000 scientists, sufficient to get congressional attention.

We called this consortium the Joint Steering Committee for Public Policy.[32] The name is replete with anonymity, a difficulty we came to appreciate when we learned that the folks on Capitol Hill very reasonably want to know exactly who you are, who you represent, and how many they might be. Scientists count publications for justification, politicians count constituents (at least those who vote, speak up, or answer polls). I dislike the former practice, but I see nothing wrong with the latter—counting constituents seems the essence of representative government.

There were close to a dozen scientists in the room for our first meeting. Virtually all were members of the prestigious U.S. National Academy of Sciences; one was to become president of that academy, another the director of the NIH; two were newly minted Nobel laureates. Several of us had become outspoken proponents for research in the news media and other public forums.[33] But virtually none of us had ever been inside a congressional office; our political inexperience was embarrassing, even irresponsible.

Faced with our inexperience and intimidated by Capitol Hill, we did the unspeakable—we hired a lobbyist. This was not well received by many of our professional colleagues, who considered it beneath the dignity of science. Hiring a lobbyist reeked of self-interest. How could we stoop so low?

But we were not deterred. What could be wrong with scientists speaking out about their "self-interest"? That quintessential San Franciscan, Ambrose Bierce, once defined politics as "the conduct of public affairs for private advantage."[34] But that can be put another way: "getting what you need from government." And we were convinced that what we needed—further support for research—was in the public interest as well as our own.

Lobbying for Science

So we hired a lobbyist, a former Democratic congressman from Maine named Peter Kyros. In retaining Peter, we obtained not only his own services, but also those of his remarkable colleague, Belle Cummins. Belle became so valued that her untimely death in 2000 evoked admiring eulogies in the scientific press and at national research meetings. The outpouring of posthumous praise and affection for Belle struck me as a singular manifestation of how much we scientists who had worked with her had changed. We had found common cause with the inner workings of government.

It should come as no surprise that Peter was in some ways our antithesis: a pragmatist who regarded the practice of government as a perfectly normal way of life. Antipodes or not, Peter and we soon reached a meeting of the minds. He grew passionate about our cause; we came to respect his savvy and effectiveness. And I soon learned that prowling the halls of Congress with a veteran can be both gratifying and entertaining, a chance to see the political personality in action, a lesson in civics unlike anything I could have imagined when suffering through my high school course in government. Anyone who has not had the experience of visiting Congress with a veteran of the place should leap at the opportunity.

Under Peter's tutelage, we formed a strategic plan. I learned far more than I expected from our tactics and their deployment. First, we wanted to get more scientists in touch with their representatives in Congress, both on Capitol Hill and in their home districts. To my surprise, we found that scientists are welcome in congressional offices, especially scientists who are constituents and come from the trenches, as opposed to polished advocates based inside the beltway. Over the ensuing years, we have engineered scores of personal visits to Capitol Hill by scientists, most of whom, like us, thereby lost their political virginity.

One essential in dealing with Congress is access. There is a method for this that must be mastered. I see nothing inherently bad in that— there are methods for all things. We learned much of the methodology from Peter. If you ever need to retain a lobbyist, here is a simple rule for efficacy: if they get you through the door, they are probably good.

But there was another, equally important lesson that we learned. Just as we need access to our legislators, we are also obliged to make our case accessible to them. The late congressman George Brown once described the problem in an interview with the *New York Times*: "[Scientists] have too great a faith in the power of common sense and reason. That's not what drives most political figures, who are concerned about emotions and the way a certain event will affect their constituency. If you are going to work in a political environment, you have to know the reasoning of the people you're dealing with. You have to talk to them realistically." Congressman Brown spoke with authority. He had a university degree in physics, had once been a practicing engineer, and throughout his political career was one of the crucial supporters of research in Congress. He also knew that talking "realistically" does not come easily to scientists, and certainly not to academicians.

My own first visit to Congress was memorable. A colleague and I were taken to see a congressman from the South who had been a member of the House of Representatives for twenty-five years. I remember that he had his feet on his desk, but that may be an embellishment from my imagination. We delivered a carefully rehearsed and deliberately brief treatise on the importance of fundamental biomedical research as sponsored by the NIH. The congressman listened patiently until we finished, then announced that we had confused him by linking fundamental research to the NIH. Why had we done this, he asked, when the NIH supports only clinical research, whereas it is the National Science Foundation that supports fundamental research?

In reality, nothing could be further from the truth.[35] A large portion of fundamental medical research is supported by the NIH. This gentleman was profoundly misinformed about how billions of federal funds were being spent every year. I winced, remembering that the congressman had by then voted on twenty-five successive federal budgets for research. Then I looked behind him, where his chief of staff was holding his head in a pronounced and deliberate display of frustration. Message: "our bosses may not always know what is going on, but we do." And they do indeed. Most of the congressional staff with whom I have dealt are bright, energetic, capable, well intentioned, and wise to the ways of the world. Think twice before arguing with any of these folks. I have tried it more than once and have generally fared poorly.

That lesson learned, my colleague and I were led down the hall to see another and very powerful member of the House of Representatives. There was a brief wait in the anteroom, during which time we were joined by a young woman who was the congressman's staff for medical affairs. When informed of our purpose, she surprised us by saying, "Sock it to him; I have been trying to change his position on this for weeks." She was as ardent about our cause as we were.

We performed again. This time, the response was more personal, more sophisticated, and more devastating. The congressman picked up a picture of his granddaughter and announced: "Gentlemen, if I do what you want, when this little girl grows up, she will have no choices left." The congressman was wary of further encumbering future federal budgets with long-term commitments—a favorite mantra in the halls of Congress and most certainly a legitimate concern.

We discussed the point for almost an hour, more time than I am usually willing to give a petitioning colleague. I was coming to understand the essence of representative government. The congressman was genuinely engaged with our issue, he had a firm and well-articulated position, he clearly loved to argue, and he was good at it.

By now, our organization has provided hundreds of scientists with opportunities like this. The opportunities arose at first as windfall. But now we have become more systematic and more strategic. In particular, we have full-time personnel organizing groups of scientist-advocates and dispatching these groups to Capitol Hill for assaults on congressional offices. At last count, we had organized in California, Illinois, North Carolina, Pennsylvania, and New York, and had hopes of expanding into New England.

A second objective of our exercise in politics was to create a nationwide team of correspondents who would generate a rapid response to crucial legislative initiatives. We soon built the membership of this team to more than two thousand, all of whom are available to be mobilized on short notice by email. The group embodies a formula that I heard early in my tuition on Capitol Hill: one letter gets a response, ten letters gets some attention, one hundred letters may get a vote. Remember that formula the next time you are trying to find the time to write to a member of Congress: you might be number one hundred.

Teaching Science to Politicians

As a third tactic, we instigated the organization of a congressional caucus on biomedical research. Caucuses are a familiar part of congressional life: self-assembled affinity groups such as the Black Caucus, the Manufacturing Caucus, the Trade Caucus, and many others meet intermittently to be briefed about their concerns and to plot legislative strategies. But there had never before been a congressional caucus devoted to research on health and disease.

We saw here the potential to create an unprecedented vehicle for the regular consideration of biomedical science on Capitol Hill. This initiative required delicate maneuvering by our lobbyist, because caucuses can be called together only by members of Congress themselves. Our role could be informal and advisory only. Several members of Congress were recruited to chair the caucus and the membership gradually grew to its current level of approximately 170.

Congressman George Gekas of Pennsylvania deserves special mention. He has been the mainstay of the caucus, attending and chairing virtually all of its meetings to date. Several years ago, the American Society for Cell Biology presented him with its award for public service. He accepted the award with an extemporaneous and impassioned address on the importance of medical research—an authentic and stirring piece of Americana that stunned the audience of jaded scientists. I have learned not to underestimate members of Congress: many are very good on a stump and bring passion to governing.

The main activity of the caucus is a regularly scheduled series of luncheon programs; there have been more than one hundred of these over the past decade. Speakers are biomedical scientists recruited from around the country with the injunction to make their remarks accessible to a general audience. The remarks are published in the *Congressional Record.* Subjects have ranged from new treatments for cancer to the way in which birds learn their songs. The objective is to inform, not to advocate. In reality, informing about the achievements of biomedical research can be the best form of advocacy.

The principal virtue of the caucus is that it provides a sustained presence for biomedical research on Capitol Hill, a means of getting science and scientists to the Hill. And we have learned that scientists care about this mission: we have by now asked well over one hundred scientists to address the caucus, and fewer than a dozen have declined. Organizers of prestigious research symposia rarely do any better.

The most important outcome of these various efforts has not been legislation, but rather civics lessons for scientists. Several hundred have visited Capitol Hill and spoken one-on-one with its denizens, many for the first time in their lives. Participants generally come away heartened by their reception and gratified that they have played a small part in representative government. The feedback we get is surprisingly grateful and idealistic, a far cry from what I had heard from that graduate student in the Moscone Conference Center. My own experience on Capitol Hill has been similarly gratifying. I rarely get what I want, but I usually come away feeling good about our government.

Some years ago, my professorial colleague and friend, Bruce Alberts, went off to Washington to become president of the National Academy of Sciences. A year or so later, he returned to San Francisco to deliver a lecture entitled "What I Have Learned in Washington." I remember the first three of his lessons vividly: there are a lot of very smart people in Washington; they work very hard; and many of them mean well.

Despite the unsavory reputation of politics, I am not convinced that, on the whole, it is any less scrupulous than other human endeavors. When science falls under close scrutiny, human shortcomings inevitably emerge. I was recently briefed on the affairs of a prominent medical institution. At the end, I remarked to a colleague that almost everything I had heard concerned conniving, calumny, cupidity, or criminality. Despite these woeful flaws, the research endeavor succeeds. So it is, in my view, with politics.

In recent years it has felt awkward to say admiring things about politics in the United States. (Then again, has it ever been otherwise?) Even those who aspired to public office often found it necessary to disparage government—a strange and harmful paradox. But the terrorist attacks of September 11, 2001, swept the slate clean. We find ourselves in urgent need of government, of the organizing and unifying force that it provides. We can once again see the importance of "[persuading] bright young people that there are issues worthy of the sacrifices a political career entails."[36] This rhetoric implies a vocational nobility easily equal to that of science, which I readily acknowledge. It also implies the drudgeries of the job, which include "being polite to strangers, compromising with idiots and reading your every unguarded remark in tabloid headlines."[37]

I once listened to a distinguished Japanese-American scientist tell the story of how he rose from humble origins to obtain a world-class education, an innovative research career, a major medical discovery, and great personal distinction—all, he emphasized, at the expense of the U.S. taxpayer and through the efforts of the U.S government. An audience of renowned scientists listened raptly as he delivered his concluding line: "God bless America."

Some years ago, the U.S. statesman Chester Bowles was asked what hope there might be for the future of peace on our planet. His response: "Fill the State Department with young people who believe fervently that peace is possible, and when they get disillusioned, get another group of enthusiastic young people to replace them."[38] The prescription of Chester Bowles is just the tonic we need for those who dismiss politics as a dirty business. The poetry of Adrienne Rich caught the spirit of that tonic:

> My heart is moved by all I cannot save:
> so much has been destroyed
>
> I have to cast my lot with those
> who age after age, perversely,
>
> with no extraordinary power,
> reconstitute the world.[39]

Few if any of us have been given extraordinary power. But we can nevertheless reconstitute the world.

A Dark Night of the Soul

Contemporary scientists generally regard the responsibility for leadership of an academic institution as a dark night of the soul. Those who assume such responsibility while still young are presumed to be otherwise inadequate, those who do so late in their careers are presumed to be otherwise superannuated. Consequently, few scientists will admit that they aspire to leadership within the university. The truth may be otherwise, of course, as pointed out by Henry Rosovsky, former dean of arts and sciences at Harvard University: "[Professors claim] to yearn for peace and quiet in the library while never missing an opportunity to engage in academic politics or games of power."[40]

During my first thirty years in San Francisco, I lived the sheltered and privileged life of a university professor. I saw myself as neither inadequate when young nor superannuated as my career approached its autumn. Yet on July 1, 1998, I became chancellor of UCSF—a move from the professoriate to an executive office, a step into the dark night of the soul.[41] Why did I let this happen?

My explanation to the local press was that "an inner voice told me this is something I should probably do." This was an honest answer, albeit spontaneously styled for the New Age public of San Francisco. But more tangible motives underlay my instinctive response.

First, I was moved by my conviction that the work of the university is the most exciting and important endeavor in the civilized world. I have never doubted this for a moment since I committed myself to an academic career more than forty years ago. In the chancellorship, I saw an opportunity to serve the fundamental purposes of UCSF more broadly than I had in the past—to assume a larger role in promoting the teaching, research, health care, and good citizenship that lie at the core of the institution.

Second, UCSF had been good to me, fostering my professional success in many and often generous ways. I had a debt to repay.

Third, I saw an opportunity to forestall a decline in my utility. In a facetious moment, I even suggested that the job might temper the shock of eventual retirement, indeed, might soon make me eager for retirement—something that I could not otherwise imagine.

Fourth, the variety inherent in the post appealed to the dilettante in me—perhaps I could finally become a Renaissance man. I was quickly disabused of that fantasy. In particular, my time for reading and reflection diminished precipitously.

Still, the decision to accept the chancellorship was not an easy one. On the one hand, I had difficulty imagining myself in the job: it held no special attraction for me—I had been conditioned to view any departure from the laboratory as a demotion, a "fall from grace," according to one member of the scientific pantheon, and I felt wholly unprepared for the transactions necessary to the position.[42] In the very week of my decision, however, I happened upon a quote from William Butler Yeats, in which he explains how the artist can exceed himself: "If we cannot imagine ourselves as different from what we are and assume the second self, we cannot impose a discipline upon ourselves . . . It is the condition of arduous full life."[43] That helped a bit: I just had to assume another self.

On the other hand, I knew that the job carried grave responsibilities and was not likely to be easy. Others have seen it the same way. Clark Kerr, an esteemed former president of the University of California, held the "strong conviction that the university's most demanding post [is] dean of a medical school or chancellor of a health science center (aside from the presidency itself)."[44] So I was forewarned, and appropriately so, I can now say. The job is indeed difficult.

Those who denigrate administrative leadership may fail to appreciate its many dimensions. First and foremost, my job as chancellor is about people, and I enjoy people immensely. Like most scientists, I am no monk—although *San Francisco Magazine* did describe me as a "lab rat," with the clear implication that this lowly species is not normally fit for chancellorial duties. (Scientists, as you will recall, see matters conversely: chancellorial duties are not fit for credible scientists.) The job of chancellor is also about politics, which I find both intriguing and necessary (which by now will come as no surprise to the reader); about education, and I am an educator at heart; about research and its representation to the general public, both of which are great passions of mine; about health care, an interest that was bred into my bone at Harvard Medical School and Massachusetts General Hospital; and about sustaining an academic community—not just an institution, but a community, and I believe heart and soul in the importance of community.

The leaders of public universities bear a special obligation to the body politic. UCSF is part of the warp and woof of daily life in San Francisco. It has grown to be the second largest employer in the city, outdone only by the city government. More than half of our fifteen thousand employees live in San Francisco proper, enriching its vitality and civic life. We have facilities in virtually every neighborhood of the city—in too many neighborhoods, some say. We operate a bus and shuttle system that carries a million riders around the city every year, with far fewer complaints than the municipal transit system—perhaps because our system is free. We gave birth to the biotechnology industry, and our faculty and their discoveries have since sired more than eighty new companies.[45] In addition, we are now building an entire new campus that will almost double our research and teaching capacity, add nine thousand jobs to the workforce, anchor a biotechnology park, and transform a derelict neighborhood—all within the city limits of San Francisco. In short, we have the potential to change neighborhoods and lives. This potential sometimes frightens people, and that fear can punish us severely (for an example, see Chapter 5). We need to exercise the potential with care and conscience; we need to mitigate the fear with education and candor.

But in turn, the body politic has an obligation to its institutions of higher learning. These are perilous times for public universities. Soaring endowments have given private universities an intimidating position in the academic marketplace. It is not presently clear whether those who are responsible for the funding of public universities have the will to keep pace. It is not even clear whether they understand what is required.

One of the fundamental difficulties in California is a chronic opposition to appropriations for university laboratory buildings.[46] The opponents argue that research is not central to the instructional mission of the university and should therefore be self-sustaining in every regard. That position is deeply antithetical to the nature of the university and its purposes: because the university is first and foremost a place of scholarship (with laboratories essential to scholarship in the natural sciences), and because the welfare and economy of our nation now depend part and parcel on the research conducted by universities, there is realistically nowhere else to turn for that benefit. Legislative testimony on behalf of research facilities for the university has become an annual ritual with great portent.

Shortly before I assumed the chancellorship of UCSF, the *New Yorker* magazine framed the expectations for the University of California in a way that would be difficult to improve:

> California in its heyday managed to make genius public property. By contrast, Massachusetts, the other great American academic enclave, has always kept genius locked away behind ivied walls. The hard question for California is whether these achievements will continue.

Twelve years on. Harold E. Varmus and J. Michael Bishop, Stockholm, December 10, 2001. (Reproduced by permission of Thomas Cech.)

> Thirty years from now, [it] is less clear to what extent . . . genius will still belong to the people of California.[47]

The university represents genius of many sorts: genius of the mind, genius of the eye and hand, genius of the heart and soul. I want to do what I can to assure that genius of every sort remains public property in California. It is for that reason, more than any other, that I entered the dark night of the soul that the chancellorship is said to be.

The opportunity to be both scholar and teacher is a privilege beyond measure. It is also a form of public service that carries many obligations, among these the search for excellence and a responsiveness to the needs of our culture. It is to choose perfection of the art over perfection of the life, and to know that a price must be paid. It is to know that "being a professor . . . remains at its core an act of conscience . . .

When smart people live at close quarters with the great mysteries (natural and human, present and past) the cost is often high; joining the knowledge business can be as risky as staring into the sun."[48]

Those of us who love the creative act must dare to stare at the sun. Amazingly enough, we are less likely to be blinded than to see more clearly. Alfred Nobel had hoped to serve this truth by establishing his prizes: "I seek not to confer distinctions for scientific achievement, but to render help where help [is] needed . . . I would like to help dreamers, as they find it difficult to get on in life."[49] We must all of us think more of helping the dreamers, wherever they can be found. In dreaming lies our salvation.

CHAPTER 3

People and Pestilence

It is now in the power of man to cause all parasitic
diseases to disappear from the world.

—*Louis Pasteur, as quoted in Edward O. Otis,*
The Great White Plague: Tuberculosis

The microbial world has been a prevailing theme in my career from the very beginning. I was lured into science by the glamour and intrigue of molecular biology, which had its inception in the study of viruses and bacteria. I first learned of molecular biology in the microbiology class taught to me during my second year in medical school. Several faculty in that course became my first icons in science. I cut my research teeth with viruses, and then, for the next two decades, pursued these objects that exist at the brink of life. A virus led me to the study of cancer cells and brought Harold Varmus to my side. Our shared study of this virus then paved the way to our Nobel Prizes. My research made me a de facto microbiologist—microbiology became my titular home in academia and the subject that I have taught throughout my academic career.

It is only natural, then, that I have steeped myself in the historical lore of the microbial world and its inquisitors. This lore is rich with colorful individuals, medical mysteries, global catastrophes, epic discoveries, and triumphs over human suffering. It spans two millennia and is growing as rapidly now as at any time in the past. It dramatizes the multitude of ways that discoveries are made. It is a story that I could not resist telling, if only in an episodic version.[1] I begin with the year in which one-third of humankind died.

The Black Death

In October 1347, Genoese trading ships arrived in the harbor of Messina, Sicily, with dying men at the oars.[2] The ships had come from the Crimea, where the Genoese maintained a lucrative trading post with contacts that extended to the farthest reaches of China. The diseased sailors had strange black swellings or "buboes" in their armpits and groins. The swellings oozed blood and pus, and were accompanied by spreading boils and bleeding into the skin. The sick died quickly and painfully. The cause of their death became known as the "bubonic plague," acquired from the bite of a flea.

As the plague spread among the populace, it assumed an even more frightful form. The sick coughed violently, spit up blood, perspired heavily, and died within hours. Everything that issued from the body —breath, sweat, blood from the buboes and lungs, bloody urine, and blood-blackened feces—gave off an unbearable stench. Bubonic plague had evolved into "pneumonic plague" and was now spreading directly from one human to another. A third form of plague, characterized by devastating internal hemorrhages (and known as "septicemic plague"), was probably also torturing the populace.

We have come to know plague in all its forms as the "Black Death." This morbid sobriquet was not coined in the Middle Ages. It originated in the scholarly literature of the sixteenth century and entered the popular lexicon in the nineteenth century. It was meant to evoke dread and catastrophe, rather than color (although hemorrhages into the skin did create black blotches on the dying).

The Black Death was not entirely new to Europe in 1347. The Athenian historian Thucydides described an epidemic of plague in the fifth century B.C., noting shrewdly that individuals who survived would not develop the disease again—perhaps the first recording of acquired immunity. The "Plague of Justinian" that ravaged Rome in 542 A.D. and helped trigger the decline of the Byzantine Empire was probably also the same disease, although some suspect instead that it was smallpox.[3] But the epidemic that began in October of 1347 had no precedent in scale or devastation.

Over six months, the plague swept across Eurasia, killing as much as one-half of the entire population between India and Ireland. "So lethal was the disease, that cases were known of persons going to bed well and dying before they woke, of doctors catching the illness at the bedside and dying before the patient. So readily did it spread that to the French physician Simon de Covino, it seemed as if one sick person could 'infect the whole world.'"[4] Vatican sources estimated deaths at over 25 million. Four-fifths of Florence perished, two-thirds of the population of Venice, half the population of Paris.

Giovanni Boccaccio witnessed the plague in Florence and included his memorable description of the epidemic in the introduction to the *Decameron*. He emphasized the profound effect that pestilence can have on the fabric of society.

> This disaster had struck such fear into the hearts of men and women, that brother abandoned brother, uncle abandoned nephew, sister left brother, and very often wife abandoned husband, and—even worse, almost unbelievable—fathers and mothers neglected to tend and care for their children as if they were not their own.
>   . . . [A] practice that was previously unheard of spread through the city: when a woman fell sick, no matter how attractive or beautiful or noble she might be, she did not mind having a manservant . . . , and she had no shame in revealing any part of her body to him . . . when necessity of her sickness required her to do so. This practice was, perhaps, . . . the cause of looser morals in the women who survived the plague.[5]

Renaissance Italy found itself besieged by repeated epidemics of the Black Death, many originating from trade with the Near East. In defense, the authorities implemented the "quarantine," a period of forty days during which ships and all their occupants remained impounded offshore—the name derives from the Italian word for forty. Although quarantine failed against the Black Death, its invention was a primitive landmark because it arose from the dim perception that the disease could be spread from one individual to another. In modern terms, the Black Death was "contagious," an "infectious disease."

Pestilence and Conquest

Two centuries after the appearance of the Black Death in Europe, Hernan Cortes landed on the coast of Mexico.[6] With him were 508 soldiers and 116 horses. Within two years, the Spaniards had laid waste to and conquered the entire Aztec civilization. The superiority of Cortes's weaponry doubtless contributed to this remarkable and despicable feat, but Cortes had an even more powerful ally—smallpox, unknowingly introduced into Mexico by the European conquistadors.

The Spaniards were themselves immune to smallpox by virtue of having survived prior infection in Europe. The immunity seemed magical to the Aztecs and added to the supernatural aura created for the conquistadors by their light complexions, horses, and weaponry. The Aztecs were able to hold their capitol city Tenochtitlan against Cortes and his army until an epidemic of smallpox struck the natives. Then the city fell quickly, eventually to become only a fabled memory buried beneath the teeming frenzy of modern Mexico City.

The conquering Europeans carried not only smallpox but also a variety of other infectious diseases into the Americas. Examples include measles, influenza, typhus, tuberculosis, whooping cough, malaria, and yellow fever. These scourges—all new to the "New World"—ravaged the Mesoamerican population, which fell from 30 million in Mexico alone at the time of Cortes's landing to fewer than 3 million only fifty years later, and to a mere 500,000 in all of North America by the eighteenth century.[7] It seems clear that without the assistance of smallpox, the Spanish victory could not have been achieved in Mexico, nor could Pizarro have accomplished his equally woeful subjugation of the Incas in Peru. It has been said that the Amerindians repaid the favor with syphilis, which may have been first encountered by the Europeans in Mesoamerica. But that idea, whatever its attractions in terms of historical justice, remains controversial.[8]

The introduction of smallpox into the Americas by the Spaniards was unintentional, and its consequences were unexpected. Arriving many years later, however, British colonists were all too aware of how they benefited from the rampages of the pox. John Winthrop, an early governor of the Massachusetts colony, remarked that "the natives, they are near all dead of smallpox, so the Lord hath cleared the title to what we possess."[9] The British also took matters into their own hands, spreading smallpox-laden blankets and handkerchiefs among the Native Americans. The architect of this strategy was the British commander-in-chief, Sir Jeffrey Amherst, namesake of the Massachusetts town that would one day be home to Emily Dickinson and give its name to the college where Harold Varmus studied. The British later used smallpox against the military forces of George Washington, in one notable instance to relieve a siege of Quebec City that might otherwise have made Canada a part of the United States.

Smallpox and the Black Death were once two of the great pestilences of humankind. Smallpox has now been eradicated from the face of the earth through human ingenuity. The Black Death lies dormant in many areas, a mere shadow of its former self, but poised to reemerge when circumstances permit. And on occasion, circumstances do permit. In September of 1994, more than 200,000 individuals fled an outbreak of plague in the Indian city of Surat. The front pages of major newspapers throughout the world covered the outbreak with daily articles, filled with anxiety that in this era of jet travel the pestilence would spread more widely. The outbreak was soon suppressed by medical intervention (and may never have been as threatening as portrayed by the press), but the panic it evoked was testimony that the six-hundred-year-old memory of the Black Death still resonates.

The subject of pestilence is large, its history glorious. The control of infectious diseases stands as one of the great achievements of civilization, wrought as much by attention to water and sewage as by the application of sophisticated vaccines and therapies. But the control is far from complete. In the face of our remedies, new adversaries have emerged—germs we previously had no idea might exist. We now often face an impasse that Boccaccio perceived in the Florentine encounter with plague: "Neither a doctor's advice nor the strength of medicine could do anything to cure this illness; on the contrary, either the nature of the illness was such that it afforded no cure, or else the doctors were so ignorant that they did not recognize its cause and, as a result, could not prescribe the proper remedy."[10]

There are further concerns. In the process of controlling infectious disease, we have disturbed a delicate balance in nature, and we are paying for this meddling in valuable coin. The decline of pestilence has helped to unleash an overgrowth of population that could one day consume society. And our promiscuous use of antibiotics has fostered the emergence of frightening microbes that are resistant to our best antidotes. None of this could have been anticipated when the underpinnings of pestilence first came to view.

Pestilence and Contagion

The contagious nature of some diseases appears to have been appreciated for as long as humankind has recorded its history. The sixteenth-century Italian philosopher and physician Girolamo Fracastoro was especially prescient.[11] In an immensely popular poem "Syphilis or the French Disease" and a subsequent essay entitled "Contagion," Fracastoro not only pointed out the transmissibility of some diseases, but also attributed the transmission to particulate "germs" that "propagate other germs precisely like themselves" and suggested that the type of germ determined the nature of the resulting disease.[12] These

Syphilis by Kathe Kollwitz, 1909. (Reproduced by permission of Bildarchiv Preussischer Kulturbesitz; © 2002 Artists Rights Society [ARS], New York/ VG Bild-Kunst, Bonn.)

are stunning ideas, anticipating modern knowledge of the microbial world by three hundred years.

The concept of contagion led to the practice of isolating afflicted individuals, first applied in earnest to lepers during biblical times, and reconceived as quarantine during the fourteenth-century epidemics of the Black Death. The earliest suspicions that some diseases might be contagious were based on anecdote, but individuals of a more experimental bent eventually provided direct evidence. The British surgeon John Hunter exemplified the experimentalist. He was known for the aphorism "Don't think, try it"—in keeping with his surgical provenance.[13] Born in 1728, Hunter was a seminal figure in the history of medicine, a brooding and idiosyncratic genius who shunned formal schooling, yet became an eminent naturalist and the foremost surgeon of his time. Aggressive and outspoken, even confessing his lethal surgical errors in publications, Hunter was both revered and reviled by contemporaries. His admirers said that he had found surgery a mechanical art and left it an experimental science. His detractors described him as a dreaming, inarticulate devotee of the scalpel, the crippled, and the corpse.

There can be no doubt that Hunter was memorable. In May of 1767, he is said to have dipped a lancet in urethral pus from an individual with overt gonorrhea and undetected syphilis, then inoculated himself by puncturing the foreskin and head of his penis. He soon developed both gonorrhea and syphilis, and concluded that the two diseases are caused by the same transmissible factor. He was correct on one count—both diseases are in fact transmissible; but wrong on the other—the two diseases are caused by very different microscopic organisms. Hunter treated his syphilitic symptoms with mercury, much as prescribed by Fracastoro two centuries before, probably to no avail: the arterial complications of advanced syphilis may well have caused his death.[14]

The claim that Hunter inoculated himself with gonorrhea and syphilis may be apocryphal.[15] It does appear, however, that he inoculated others with material from syphilitic lesions in his efforts to understand the genesis of the disease—in experiments that would be egregious violations of modern medical ethics. Hunter became an outspoken authority on venereal disease, with attitudes toward its prevention that were enlightened for his time. Later in his career, he provided a suitable counterpoise for his experiments with venereal disease by performing the first successful artificial insemination of humans (infertility is one of the long-term outcomes of gonorrhea in women). He is entombed at Westminster Abbey "in grateful veneration for his services to mankind as the founder of scientific surgery."

The world was slow to exploit Hunter's revelations about venereal disease. Elie Metchnikoff was the Russian bacteriologist who discovered phagocytosis, the defensive engulfment of alien objects such as microbes by certain blood cells (work for which he received the Nobel Prize in 1908). He was also a careful student of syphilis who preached the importance of what we now call "safe sex." In a lecture delivered in 1906, he complained that "in spite of having been adequately warned, young doctors and students of medicine furnish a large portion of victims of syphilis. Ignorance is therefore not the only cause."[16] Nor, it seems, does ignorance necessarily engender bliss.

A century after John Hunter, the Viennese physician Ignaz Semmelweis became convinced that a deadly infection of women immediately following parturition was caused by the carelessness of physicians and students, who came to the delivery room directly from autopsies without so much as rinsing their hands. The disease was "puerperal fever" (or more popularly, "childbed fever"), which we now know to be caused by the infectious bacterium streptococcus. The frequency of puerperal fever in obstetrical wards at the time was dreadful: as many as one in every three postpartum women died of the disease. But the deaths were occurring mainly in those wards attended by physicians and students. Births handled by midwives had a very low incidence of puerperal fever, as did unattended births at home or even in the streets. From these circumstances, Semmelweis deduced that students and physicians were carrying "cadaveric particles" from the autopsy room to the laboring woman, and the result was puerperal fever. Thirty years later, Louis Pasteur would reveal those "cadaveric particles" as the bacterium streptococcus, pointing out in particular their propensity to form long chains of microbes that assist in their identification.

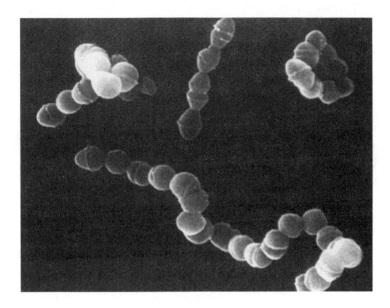

Streptococci. Electron micrograph by David M. Phillips. The bacteria have propagated into the characteristic chains originally described by Louis Pasteur. (Reproduced by permission of the *New England Journal of Medicine* and David M. Phillips.)

Almost a century before Semmelweiss, British physicians were writing of the possibility that an external agent might cause puerperal fever. But the ideas were ill-formed and came to naught until Semmelweiss's time, when some of his contemporaries in Britain began to advocate strict hygiene as a prevention for puerperal fever. It was Semmelweiss who provided the clinical data that made the case sound.

Semmelweis supervised one of the largest maternity services in the world at the time, so he was in an excellent position to test his theory. He compelled the students and physicians under his charge to wash their hands with dilute chlorine and the incidence of puerperal fever dropped dramatically. But many of his peers resisted his conclusions, in part because he was very slow to publish his findings, and in part because his humble origins in Hungary and his own deep sense of social inferiority had impeded his access to the medical establishment in Vienna—he had been denied positions in pathology and medicine, and only then had turned to obstetrics, which was viewed as a lowly pursuit for physicians.[17]

Undeniably, Semmelweis was a difficult man. Here is what he wrote in an open letter to one of his detractors:

> Herr Professor, there remains nothing else but to adopt my doctrine, if you still want to salvage something of your reputation, whatever of it is still left to salvage. If you continue to adhere to [false] doctrine, your reputation will disappear from the face of the earth . . . Herr Professor has proven that in spite of a new lying-in hospital furnished with the best equipment, a great deal of homicide can be committed, if only one possesses the necessary talents.[18]

It is easy to be sympathetic with Semmelweis's vehemence. He was opposed with bigotry, obstinacy, foolishness, and cupidity. And he was a man on a mission:

> My Doctrine is not established in order that the book expounding it may molder in the dust of a library: my doctrine has a mission, and that is to bring blessings into practical social life. My Doctrine is produced in order that it may be disseminated by teachers of midwifery until all who practice medicine, down to the last village doctor and the last village midwife, may act according to its principles; my Doctrine is produced in order to banish terror from the lying-in hospitals, to preserve the wife to the husband, the mother to the child.[19]

Rejected by the academic elite of Vienna, Semmelweis retreated to the intellectual backwater of Pest (later amalgamated into Budapest), Hungary, where he once again banished puerperal fever from the maternity wards by his insistence on hygiene. Near the end of his career, Semmelweis finally brought his arguments together in a ponderous book that first reviewed all of the available evidence for his views on puerperal fever and then railed slanderously against his detractors.[20] It did nothing to redeem his reputation. Over the last years of his life, Semmelweis drifted into psychosis, possibly suffering from Alzheimer's disease. He was confined to an asylum and apparently beaten to death by attendants in efforts to constrain him.[21]

Semmelweis's distressed life and sad end stand in stark contrast to his epochal achievements. He was a pioneer in the collection and rigorous analysis of clinical data. It can easily be said that he was among the founders of clinical research as we know it today. The principles that he enunciated and authenticated saved innumerable lives and eventually transformed obstetrical practice. Toward the end of his career, he applied the same principles to gynecological surgery, thus taking the first step toward the sterile practices that are now routine in all operating rooms. In the words of one biographer, he was "a martyr to the world's stupidity, . . . one of the great tragic figures of all history."[22]

America's own Oliver Wendell Holmes also figured out the origin of puerperal fever, some years in advance of Semmelweis. Holmes had no compunctions about publication or forceful argument: "The disease known as Puerperal Fever is so far contagious as to be frequently carried from patient to patient by physicians and nurses."[23] Acting on this belief, Holmes was able to greatly reduce the incidence of puerperal fever in the maternity wards of Boston. But like Semmelweis, Holmes saw his doctrine resisted and ignored. At the time, neither man was aware of the other's work, although Holmes later made guarded reference to secondhand reports of Semmelweis's claims.

One day after the death of Semmelweis, on August 12, 1865, the Glasgow surgeon Joseph Lister applied the first antiseptic dressing in history, to a compound fracture of the tibia of an eleven-year-old boy who had been run over by a cart. Lister used carbolic acid for antisepsis, and the results exhilarated all observers. The wound healed without so much as a hint of the otherwise inevitable infection.[24]

In due course, Lister applied antisepsis to surgery of all sorts. His successes soon made him a legend. Before Lister, postoperative wards were cesspools of gangrenous wounds that emitted a nauseating stench that clung to the surgeon's clothes at the dinner table and caused mortality rates as high as 75 percent. After Lister, surgeons knew the enemy for the first time and could take measures to defeat it, particularly aseptic surgical procedures aimed at preventing infection—procedures clearly preferable to the antisepsis that Lister had devised to eliminate infection once it had occurred. Ignaz Semmelweis had been vindicated and the modern era of microbiology had begun.

Lister triumphed where Semmelweis had failed because of two factors. The first was his personality: he was a saintly man who left his

Quaker faith in order to marry the woman he loved, but retained the missionary zeal required to prevail over the initial doubts of his peers. The second factor was Lister's scholarly nature: he was a careful student of the scientific literature who seized early the emerging evidence that contagious disease is caused by living microorganisms and wasted no time in alerting the world to his results.

Microbes Uncovered

Microorganisms were discovered late in the seventeenth century by Antoni van Leeuwenhoek, a Dutch draper who became fascinated with the recently invented microscope, mastered both its fabrication and use, and used it to examine virtually anything at hand—from slime to semen. Leeuwenhoek's written records and drawings suggest that he succeeded in identifying all of the major forms of microbes except viruses, which cannot be seen with an ordinary microscope. He even recorded the first description of what we now call *Giardia*, after examining his own diarrhea. Leeuwenhoek transmitted his drawings to the Royal Society of London for dissemination and preservation, thereby creating one of the most elegant visual legacies in biomedical science.

Remarkably, the connection between Leeuwenhoek's discoveries and the genesis of infectious disease was not made until a century later. The earliest connections were made in modest ways. The first microbial cause of disease to be identified was a fungus of silkworm, and the first microbe to be implicated in a human disease was a fungus of the skin. But then two remarkable men entered the picture and the pace quickened.

One was Louis Pasteur, an austere but passionately intelligent Frenchman whose scientific writings bristle with imagination and vision—they make good reading to this day. The other was Robert Koch, a dour and acid German who often disputed Pasteur's views, usually without success. Between these two men, microbiology as we know it now was born, and the quality of human life was transformed.

Pasteur ranged widely over the problems of infectious disease and microbiology. He described the microbial causes of fermentation;

saved the French wine and silk industries from death by infestation; developed the procedure of "Pasteurization" for the decontamination of foodstuffs; refuted once and for all the doctrine of spontaneous generation of life; and became a public idol when he developed vaccines for several infectious diseases, including the dreaded rabies.

The career of Louis Pasteur offers two lessons that are of particular note to scientists. First, Pasteur was a chemist, not a physician. His intrusion into the realm of microbiology was not well received by the French medical establishment, yet his discoveries helped transform the practice of medicine. Second, many of Pasteur's insights arose from efforts to solve practical problems, such as threats to the beverage and silk industries, and infections encountered in veterinary medicine. He was oblivious to any distinction between fundamental and applied research (or at least to the relative cachet of the two).

Louis Pasteur was no saint. He was at the very least self-absorbed and ill-tempered. He is also reputed to have pilfered other scientists' ideas and materials, and to have misrepresented his experiments when doing so suited his purposes. But he remains a figure for the ages. Few individuals have had as great an effect on the health and welfare of humankind.[25]

Pasteur himself had no doubt about his greatness. At the age of twenty-nine, he consoled his neglected wife with the assurance that his work would "lead her to posterity."[26] She apparently bought the argument. The next year, she wrote her impatient and skeptical father that "the experiments [Louis] is undertaking this year should give us a Newton or Galileo if they succeed."[27] At the time, Pasteur was in pursuit of a "cosmic asymmetric force" that might account for all of life. He never found this force. But his subsequent work on microbes did indeed make him the equal of Newton or Galileo.

Robert Koch began his career as an idealistic and unassuming country physician who, through his great successes as a scientist, grew into what his American biographer called a "crusty and opinionated tyrant."[28] But do not be deceived: passion runs deep. In 1889, Koch met Hedwig Freiberg. He was then forty-seven, she seventeen. They fell deeply in love and, three years later, married one week after the final decree of Koch's divorce from his first wife of twenty-six years, Emmy

Fraatz. Emmy was unjustly served. Working as an unordained nurse-practitioner, she had maintained Koch's private medical practice while he pursued the work that was to gain him lasting fame. In her words, "It was my job to find out how sick a patient really was, and to send away those who didn't really need medical attention."[29]

Koch wrote to Hedwig, while their relationship was still adulterous: "Dearest Hedchen, if you love me, then I can put up with anything, even failure. Don't leave me now, your love is my comfort and the beacon that guides my path."[30] Imagine that, from a crusty and opinionated tyrant. The marriage endured until Koch's death at the age of sixty-seven in 1910. Hedwig never remarried. She became a devoted student of Eastern religions and lived until 1945.

Koch began his pathbreaking work while practicing as a physician in Wollstein, a town of three thousand, situated in the geographical limbo between Germany and Poland. (The house in which he worked there now bears a memorial plaque. It has, at various times, been worded in either German or Polish, according to the fortunes of war.) Koch was remembered as a popular and capable physician. But he had additional aspirations. Using personal income to finance a primitive laboratory, Koch began to pursue the microbial causes of infectious disease. His results quickly gained him an appointment at a newly established state laboratory in Berlin, to which he moved on three days' notice (Koch was not one to waste time—his motto was *nunquam otiosus*, "never idle"). He was soon enjoying a meteoric rise to international fame.

Koch built his career on the development of techniques for the identification and isolation of microbes. The laboratory isolation and propagation of microbes was first performed in liquid medium that provided the requisite nutrients. It was difficult (albeit not impossible) with this procedure to separate one microbe from another. But in 1875, Joseph Schroeter reported the growth of bacteria as isolated colonies on the surface of cut potatoes, and later, on medium solidified with starch paste.

Koch first modified the paste by using gelatin rather than starch. But gelatin melts at temperatures that are optimal for the growth of most bacteria. So Koch substituted a substance derived from seaweed

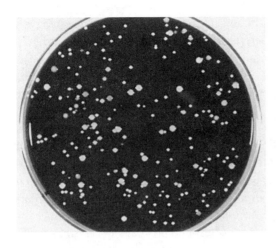

Bacterial colonies. Individual colonies of staphylocci, propagated on an agar plate. Each white mound contains close to a billion bacteria.

known as agar, allowing the medium to solidify in flat round dishes introduced by his colleague, Julius Petri (for whom the dishes are now named). The idea of using agar had come from Walter Hesse, who had worked briefly with Koch. Hesse, in turn, had been inspired by his wife, Fanny, who used agar in making jams and was familiar with its coagulative properties.[31]

Thus was born the "agar plate," the most familiar implement of modern microbiology. Myriad agar plates are used every day in the diagnosis and management of infectious diseases. Clinical specimens such as throat swabs, sputum, or blood are spread over the agar surface and then incubated at body temperature. Within a day or two, microbes present in the specimen propagate into small, visible piles known as "colonies," containing millions of organisms. If the colonies are adequately separated from one another, each represents the progeny of a single microbe and is therefore genetically pure. This purity can be essential to identification of the organism and determination of which antibiotics might inhibit its growth. Much of this would have sounded abstruse to general readers in the not too distant past. But it has been made commonplace in the United States by the intense press coverage of bioterrorist attacks with anthrax.

Koch used the agar plate to perfect procedures for isolating and propagating pure specimens of a microbe. Many consider this the most important accomplishment in all the history of microbiology: it allowed the detailed characterization of individual microbes and provided the means to implicate specific microbes as causes of disease. Koch used his procedures to identify the causative agent of anthrax, the agent of human cholera, and—in his crowning achievement—the organism that causes tuberculosis, *Mycobacterium tuberculosis*.

Anthrax deserves special note. Once familiar only to farmers and ranchers as a disease of livestock, it is now a household word as a result of its use in bioterrorism through the U.S. mail. It may have been a co-conspirator in causing the Black Death. The Black Death has long been attributed to a germ known informally as the plague bacillus (the latter word means "rod" and denotes the shape of the offending microbe when visualized with a microscope), and formally as *Yersinia pestis*. But some scholars have argued that the rapid spread of plague through rural areas is difficult to ascribe to plague bacillus and may, instead, have been due to anthrax, acquired from cattle.[32] Yet another contemporary hypothesis attributes the Black Death to Ebola virus or one of its kin.[33] Neither the proponents of anthrax nor those of Ebola virus dispute the role of *Yersinia pestis* in the various forms of plague as we now know them. But the ambiguities of the historical record do leave open the possibility that the catastrophic epidemics in the Middle Ages had more than one cause.

Anthrax was the source of conjoint triumphs for Robert Koch and Louis Pasteur, despite their bitter rivalry: it was the first microbe isolated by Koch with the techniques that he had pioneered, and it provided Pasteur with a major success in immunization (he used sheep for the demonstration)—a highly publicized episode that contributed greatly to his celebrity.[34] While the isolation of the anthrax microbe was a genuinely original discovery, Koch's isolation of the cholera agent was a reinvention of sorts. Thirty years earlier, an Italian physician named Filippo Pacini had first reported the microscopic detection of a curved bacterium in intestinal tissue affected by cholera. Pacini suggested that this microbe caused the disease. Koch did him one better by isolating the organism, propagating it in pure form, and demonstrating its association with cholera. In a bow to priority, the official name of the bacterium was revised in 1965 to be *Vibrio*

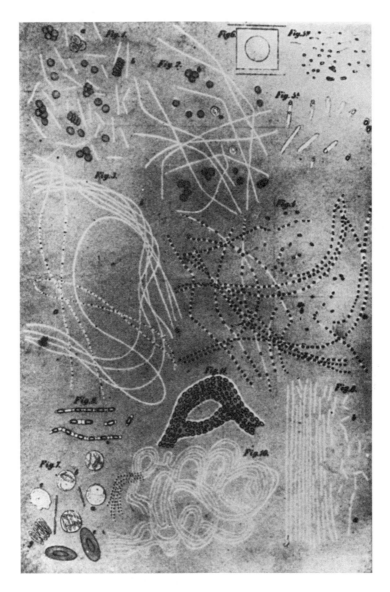

Anthrax. A historic set of drawings that illustrates the growth of anthrax bacillus in various forms and settings. The original was a color lithograph, published in Robert Koch, "Die Aetiologie der Milzbrand-Krankheit, begrundet auf die Entwicklungsgeschichte des Bacillus anthracis," *Beitrage zur Biologie der Pflanzen* 2 (1876): 277–310. It was in this publication that Koch rigorously identified the bacterial cause of anthrax, the first demonstration of a specific association between a particular bacterium and a particular disease.

cholerae Pacini. But it was Koch who actually laid hands on the culprit and decisively incriminated it as the cause of cholera.

Koch was not devoid of mercenary instincts. At about the time of his divorce, he developed an extract of the tuberculosis germ that he thought might work as a vaccine. Needing money to finance his divorce and remarriage, Koch kept the formulation secret in the hope of marketing the material at considerable profit. There was to be no profit. The extract failed as a vaccine, although it is still used in the familiar skin test that is used to detect previous infection with *Mycobacterium tuberculosis*. Then as now, diagnostics were less lucrative than therapeutics. So Koch had to find other means to finance his personal upheavals.

Koch was every bit as self-assured as Pasteur. Greeted with great honor during a visit to New York City, Koch responded: "If I think of all the praise which you have heaped upon me, I must, of course, immediately ask myself if I deserve it. Am I really entitled to such homage? I guess that I can, with a clear conscience, accept much of the praise you have bestowed upon me."[35]

Koch was twenty-one years younger than Pasteur and never felt appreciated by the elder titan. He retaliated with professional venom: "Of these conclusions of Pasteur on the etiology of anthrax, there is little which is new, and that which is new is erroneous . . . Up to now, Pasteur's work on anthrax has led to nothing."[36] It led in fact to a vaccine that was used successfully to immunize farm animals throughout the world until supplanted years later by a more advanced preparation. Neither scientist published in the other's language, so the vitriolic dispute that grew up between them was byzantine in the extreme. Pasteur advocated the use of vaccines to control infectious disease, whereas Koch was a pioneer in sanitation and had nothing good to say about vaccination. It was a senseless debate. Modern medical practice utilizes both approaches to good effect.

Pasteur and Koch were never reconciled. In retrospect, much of their discord seems to have been rooted in the strong antipathies between Germany and France that were to culminate in the First World War. Both Koch and Pasteur expressed such antipathies in their correspondence with other scientists, but these founding fathers of microbiology never wrote to one another. In 1905, Koch received the Nobel Prize in Physiology or Medicine for his discovery of *Mycobacterium tuberculosis*. (Pasteur had died six years before the prize was established.) Koch complained about the strenuous trip to Stockholm, but, by his own account, Hedwig had a great time.

Microbes in Brief

The general public thinks of all contagious creatures that cause disease as "germs." In most instances, either bacteria or viruses are actually to blame. But there is a bit more to the story than that. We now know that microorganisms take five major forms: protozoa (such as the amoebae that inflict dysentery and worse on many an intrepid traveler), fungi (as in bread mold or the thrush familiar to most mothers), algae (familiar as the green sludge floating in stagnant ponds and the source of "red tide"), bacteria (pus-laden boils may represent our most common cause for complaint against these creatures, but anthrax has the greatest currency), and viruses (think of the common cold or influenza).

All infectious microbes are individual cells with diameters no greater than one millimeter. They can been seen only with the aid of a

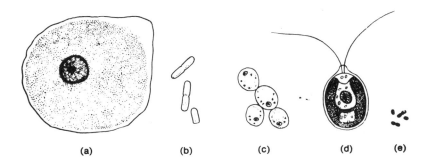

(a)   (b)   (c)   (d)   (e)

Microbes. The various forms of microbes are illustrated at a magnification of approximately 10,000-fold. From left to right: (a) amoeba, (b) large bacterium, (c) yeast, (d) alga, (e) small bacterium. (Adapted from *Microbial World*, Stanier/Doudoroff. Reproduced by permission of Pearson Education, Inc., Upper Saddle River, N.J.)

microscope, or in the special case of viruses, only with the exceptional magnification provided by an electron microscope. These units never assemble into communities of cells with distinctive functions, such as the different tissues found in mammals (in formal nomenclature, microbes are said to be "unicellular"). All microbes except algae include variants that cause infectious disease, and even algae can cause us grief through toxins released into water—the familiar "red tide."

At least some veterans of high school biology can dimly recall another form of taxonomic distinction among microbes: prokaryotes and eukaryotes. Prokaryotes comprise all bacteria; eukaryotes, all remaining forms of microbes except viruses. The fundamental difference between prokaryotes and eukaryotes is that the genes of eukaryotes are contained within a membranous housing called the nucleus. Viruses are neither prokaryote nor eukaryote, but simpler and much smaller forms, whose structure and mode of replication place them at the very edge of life—they are incapable of independent existence and must penetrate cells in order to replicate.

Microbes reproduce with astonishing speed: one million times more rapidly than we humans. The rapidity of microbial growth has far-reaching consequences. The genes of microbes change (or "mutate") at conventional frequencies, with every gene suffering a mutation within the span of one million cell doublings, and a million comes quickly in a population of bacteria that is doubling in number every twenty minutes. In addition, many microbes shuttle genes back and forth among themselves, by direct transfer or by portage in viruses.

By means of mutation and shuttling, new genetic variants accumulate relentlessly. The genetic plasticity of microbial populations has produced a bewildering diversity. Modern techniques can detect as many as ten thousand different types of bacteria in a single gram of soil. Estimates of the total number of bacterial species run as high as ten million. Insects run a distant second, with far fewer species. Life is indeed "dominated by its bacterial mode."[37]

Rapid microbial growth allows new genetic variants to expand their population immensely over brief periods of time, particularly under conditions of selection that favor the survival of the new variant (such as the presence of a hostile antibiotic or antibody to which only the variant is resistant). Hence, antibiotic resistance can emerge from even undetectable beginnings to dominate rapidly a population of bacteria—physicians know the frustration and threat of this occurrence as an everyday event. By the same token, microbes can evolve rapidly to elude the immune response—witness the annual need for new vaccines against influenza virus.

Friendly Microbes

I have thus far done microbes an injustice, because I have given the impression that they are inevitable enemies. Nothing could be farther from the truth. The overwhelming majority of microbes, numbers beyond reckoning, are harmless to us; and many of them are vital, or at least important to our comfort, convenience, pleasure, and even indulgence. These realities were resisted by the medical community well into the twentieth century. In the traditional view, microbes were categorically bad. The facts tell a very different story.

Algae of the sea constitute the largest single source of photosynthesis on the globe, remove 10 billion tons of carbon dioxide from the atmosphere every year (thus combating global warming), and are an essential ingredient in the food chain of the sea. Bacteria and other microbes efficiently decompose the organic debris from higher organisms, helping to save us from burial beneath a vast waste heap. Yeast carry out various forms of fermentation that enhance the quality of our lives (think of bread, beer, and wine), as Pasteur was first to show persuasively. Bacteria assimilate nitrogen from the soil for plants and synthesize nutrients for us and other beasts. Fungi offer themselves up to us as mushrooms. And in a twist of natural irony, both bacteria and fungi provide us with antibiotics that are weapons against their own kind and other microbes as well.

We ourselves are inhabited by microbes, known as our "normal flora."[38] Their number is extraordinary—one hundred trillion bacteria on each of us. They have located themselves in every niche and cranny of our body's surface, and they teem within our intestines—the fecal mass is composed largely of bacteria. They appear not to be absolutely essential for our survival, but they are very useful. Perhaps their most important service is uncommonly subtle: their very presence keeps away potential enemies, precluding the easy colonization of our bodies by new, possibly unwelcome, "transient" microbes. One telling example of this service goes as follows. Defecation deposits millions of intestinal bacteria on our perianal skin, our best efforts with toilet paper notwithstanding. Within hours, however, these interlopers are gone, scavenged by the microbes who call the perianal skin their permanent residence. In other words, our normal flora are happy where they are, they do not yield the field easily, and thus, we are better off with them than without.

The conclusion from all this is that the eradication of our normal flora is an unnatural act that promises and yields unhappy consequences. Physicians are obliged to remember this point every time they are tempted to prescribe an antibiotic without good cause, because antibiotics make no distinction between normal flora and pathogen. One familiar consequence is the diarrhea that often accompanies the oral administration of an antibiotic and arises from elimination of bacteria that normally dwell in our intestinal tract. In the United States, the most common impetus to unjustified use of antibiotics in medical practice is probably the quixotic desire to achieve a quick fix for the common cold—alas, the viruses that cause most upper respiratory infections are not affected by antibiotics. In many areas outside the United States, the problem is compounded by the availability of antibiotics without prescription—there the instinct for remedy can run rampant.

W. H. Auden wrote verse to our normal flora:

> For creatures your size I offer
> a free choice of habitat,
> so settle yourselves in the zone
> that suits you best, in the pools
> of my pores or the tropical forests
> of arm-pit and crotch,

in the deserts of my fore-arms,
or the cool woods of my scalp

Build colonies: I will supply
adequate warmth and moisture,
the sebum and lipids you need,
on condition you never
do me annoy with your presence,
but behave as good guests should
not rioting into acne
or athlete's-foot or a boil.[39]

Auden had it right. The happy affiliation of microbe and human sometimes goes awry. We have developed a small vocabulary to describe the circumstances of this mishap. Any organism that gains nourishment at the expense of another (the host) is known as a parasite. Parasitism may be of no consequence (saprophytic), of mutual benefit to microbe and host (symbiotic), or deleterious to the host (pathogenic). A host to a parasite is, by definition, infected, but the infected host may not be ill. The distinction between infection and disease involves some of the most fundamental puzzles in the study of microbes.

Infection and Disease

Infection without disease is the rule in nature. This fact comes as a surprise to many. But it is true. We are infected by our normal flora without harm, and with rare exceptions (such as HIV) even infections with dangerous pathogens may cause disease infrequently. We know very little about why some individuals develop disease in response to infection with a particular microbe, whereas others do not, but variations in our own genes appear to have a great deal to do with it.

The idea that infection with a pathogenic microbe necessarily leads to disease is a deeply rooted misconception. For example, those iconoclasts who claim that HIV does not cause AIDS are fond of citing the fact that not everyone who becomes infected with the virus develops disease. In reality, HIV is among the most efficient pathogens known to medical science, but there are a fortunate few who escape the ravages of AIDS despite infection with the virus, for reasons that are only partially known.

A small vocabulary of formal terms is used to describe the ability of microbes to cause disease. All of the terms derive from the noun "pathology," which refers both to abnormalities caused by disease and to the study of those abnormalities. The derivative family includes "pathogen" (any microbe that can cause disease), "pathogenesis" (the process of producing disease), and "pathogenicity" (the ability to cause disease). Infection and disease occur in two distinctive but merging patterns. They may be concentrated in time and space and thus known as "epidemics"; or they may be present in an area or population continuously, in which case they are called "endemic." The term "pandemic" refers to a particularly widespread, generally global epidemic.

The relative effectiveness of a microbe as a pathogen is defined as its virulence. Some microbes are highly virulent: they cause disease in a majority of infections. Others may cause little or no trouble and are known as avirulent. These terms are sometimes used in other ways, most commonly to describe the relative severity of disease that follows infection. The context of usage determines the meaning, the sort of ambiguity that would not usually be welcomed in science or medicine.

The mechanisms of pathogenicity and virulence are poorly understood. To sample the complexities of the story, however, consider the story of Max von Pettenkofer—a distinguished Bavarian physician who was active in the late nineteenth century. Von Pettenkofer was the source of many of the modern ideas of hygiene and was responsible for providing Munich with a pure supply of water. Yet he never accepted the role of contaminated drinking water in the transmission of either typhoid fever or cholera. He did not believe that bacteria alone could cause disease, instead invoking a combination of factors, including a mysterious miasma derived from the soil.

To prove his point, in 1892, at the age of seventy-four, von Pettenkofer drank a hefty dose of cholera bacteria that had been isolated from the stools of an individual who had died of cholera. He was joined in this novel ingestion by a number of his colleagues, including Elie Metchnikoff of phagocytosis fame. Several of these intrepid individuals experienced abdominal pain and diarrhea, and all had prodigious amounts of cholera bacteria in their feces for many days. Since none of them became seriously ill, however, their ailment was dismissed as not being cholera.[40]

Pyotr Ilich Tchaikovsky did not fare as well under similar circumstances. In the fall of 1893, one year after von Pettenkofer's self-experimentation, an outbreak of cholera occurred in Tchaikovsky's home city, St. Petersburg, Russia. On November 2, he drank a tumbler full of water that had not been boiled. He died four days later of cholera, only nine days after conducting the premier of his sixth symphony, an especially gloomy piece of music that foreshadowed what was coming. Program notes in concert halls sometimes suggest that Tchaikovsky had actually drunk poison when threatened with disclosure of a homosexual relationship. This claim appears to be discredited by available evidence. Whether Tchaikovsky exposed himself deliberately to the cholera bacillus instead, we may never know.

Von Pettenkofer had obtained the cholera bacteria for his experiment from none other than Robert Koch and wrote to him derisively afterward: "Herr Doctor Pettenkofer presents his compliments to Herr Doctor Professor Koch and thanks him for the flask containing the so-called cholera vibrios [the bacteria whose role in cholera Pettenkofer had denied], which he was kind enough to send. Herr Doctor Pettenkofer has now drunk the entire contents and is happy to be able to inform Herr Doctor Professor Koch that he remains in his usual good health."[41] Pettenkofer's taunt failed to mention either his diarrhea or the fact that he had suffered a bout of cholera some years before. So it is likely that he possessed at last partial immunity against the disease.

Von Pettenkofer later wrote the following about his experiment: "Even if I had deceived myself and the experiment endangered my life, I would have looked Death quietly in the eye for mine would have been no foolish or cowardly suicide; I would have died in the service of science like a soldier on the field of honor."[42] Von Pettenkofer had deceived himself, of course. Cholera is indeed caused by the bacterium he swallowed. The truth eventually became inescapable. In 1901, von

Pettenkofer committed suicide by a gunshot to the head. Some of his acquaintances claimed that he had finally realized the error in his opposition to Koch and considered himself academically disgraced. But it is also possible that his brain was addled for other reasons, and that he went to his grave believing in his own mistaken theories about cholera.

The Origins and Evolution of Pathogenesis

How do pathogens arise? Why does nature tolerate delinquent microbes? The answers to these questions are obscured by great lengths of evolutionary time and by our relative ignorance of the factors that determine pathogenicity, but four principles appear to be valid.

First, pathogens are never unique in the microbial world. Either they are errant relatives of harmless microbes, often of our normal flora; or they can be found in one species as a harmless infection, but cause trouble when transferred to another, "accidental" host. The Black Death originated when the causative bacterium was transferred from wild rodents to humankind. Circumstantial evidence traces the origins of measles, smallpox, tuberculosis, influenza, whooping cough, and malaria to various domesticated animals.[43] And that most contemporary of pathogens, HIV, seems to have made its way sometime in the not too distant past from chimpanzees to humans in Africa, perhaps by exposure to the blood of infected animals (chimps are used by humans as food).

Second, humankind got itself in trouble with the microbial world when it began to domesticate animals and to live at close quarters. Many of our most deadly infectious diseases were apparently acquired from newly tamed animals, then perpetuated through chains of infection made possible by the crowding in village settlements.

Third, pathogenicity is sometimes conferred on a microbe by an invader. For example, the gene that gives rise to cholera (by producing a toxin) was carried into the cholera bacterium by a virus, then stabilized by assimilation into the genetic fabric of the bacterial host. In such a circumstance, the assumption of pathogenicity can be abrupt and catastrophic, having bypassed the more sedate pace of natural selection—the pathogen is formed immediately upon receipt of the necessary gene(s) and can then disseminate rapidly and widely.

Fourth, pathogenicity may offer a benefit to microbes. Many of the adverse consequences of microbial infection appear to facilitate transmission of microbes from one host to another. Diarrhea can assist fecal spread (cholera); respiratory symptoms, aerosol spread (influenza); skin sores, spread by contact (various venereal infections). But there are exceptions to this formula that offer less justification for pathogenicity. For example, microbes transmitted by insect vectors generally fail to elicit manifestations that might facilitate spread, yet can be ferociously pathogenic. Such is the case for malaria, yellow fever, encephalitis, and many others.

The death of an infected host is neither necessary nor even desirable from the vantage point of the microbe. The longer a host lives following infection, the greater the opportunities for microbial propagation and dissemination. So it stands to reason that, over time, natural selection might moderate the consequences of infection for the host.[44] As an example, consider the treponematoses, diseases that include pinta, yaws, bejel, and syphilis. All four are caused by closely related bacteria known as spirochaetes.

The agent that causes pinta is thought to be the most ancient of these microbes, and it causes the least severe disease. Each successor in the evolutionary scheme is a more vigorous pathogen, syphilis being the "youngest" and the most severe. Since humans are the sole hosts for these bacteria, it appears that sustained cohabitation of microbe and host eventually moderated the consequences of infection. Such moderation is evident even in the history of syphilis itself, which was described as a much more fulminant disease in sixteenth-century Europe than at present (although prudence still dictates that syphilis be avoided—its long-term consequences can be fearsome).

It was once thought that all parasites would become less virulent to their hosts over time. But this view is now in question. There are circumstances in which an increase of virulence might be favorable for certain microbes. For example, if illness impairs transmission, then the causative microbe will probably evolve to lesser virulence. But if as suggested before, the debilities caused by infection happen to favor transmission, then virulence may persist or even increase. There are uncertainties in these formulations, however, that dramatize the hazards of teleological reasoning. It is possible to rationalize virtually any change in virulence—for the worse or for the better—as a response to natural selection. Having accepted the logic of evolution, biologists are sometimes too facile in its application.

Virulence is not the only determinant of microbial transmission. The stability and resilience of microbes also have a say in the matter. The more delicate a microbe, the more likely it is to be transmitted by direct means, such as personal contact, inhalation, or insect vectors. The treponematoses exemplify such fragile microbes. In contrast, more robust microbes like poliovirus can survive prolonged periods outside a living organism. The syphilis bacterium is both delicate and fastidious, requiring the warmth and moisture of the genitourinary tract to survive. In contrast, poliovirus is remarkably stable, having been found in archaeological specimens that are many centuries old—it is transmitted in human feces and readily survives prolonged residence in soil, sewage, and even chlorinated swimming pools.

Watching a Virus Evolve

Human meddling with ecology has provided a dramatic and carefully analyzed example of how microbes can evolve. Until 1859 Australia had no rabbits. In that year, Thomas Austin released one dozen wild European rabbits on his property near Geelong in Victoria. The rabbits thrived and spread like the plague, since they encountered no natural predator in Australia. By the turn of the century, the immigrant rabbit had become the number one pest of Australian farmers. In desperation, the government of New South Wales solicited proposals for how this problem might be solved. One submission came from none other than Louis Pasteur. It was deemed unacceptable (one of Pasteur's few failures in search of resources).

Nothing better came along until 1950. Then the Australian Commonwealth Quarantine Department turned to biological warfare, introducing myxomatosis virus into the wild rabbit population. Myxomatosis virus lives in quiet concert with the Brazilian rabbit, causing no disease and persisting effectively in the wild rabbit population. To the European rabbit, however, myxomatosis virus is a deadly pathogen. The strain introduced into Australia in 1950 was known to kill 99 percent of all infected European rabbits in a matter of a few days after infection.

Transmitted by mosquitoes, the virus spread over an area greater than that of Western Europe within three months. The rabbit population was ravaged. Australian farmers rejoiced in the prospect of extinguishing their worst pest, while Australian scientists followed with morbid fascination this epidemic of a foreign pathogen in a previously unexposed host population.

Then the epidemic reached Mildura on the Murray River, in the northwest corner of Victoria, and uproar followed. At almost the same time as the rabbits of Mildura began to show symptoms and die, a dozen cases of severe encephalitis appeared in local children. There had been no such disease in that region for a quarter of a century. The encephalitis spread among the population, and suspicion grew that a virus liberated by the public authorities to kill rabbits was now killing children.

Into this nasty breech stepped two of Australia's most prestigious scientists, Frank Fenner and MacFarlane Burnett. They and other of the country's microbiologists had been chasing the cause of the epidemic encephalitis since its inception. They knew from their data that the human disease could not be caused by myxomatosis virus.

To dramatize their point, Fenner and Burnett called a press conference and there injected themselves with enough myxomatosis virus to kill one hundred rabbits. They suffered no adverse consequences, to the relief of numerous bureaucrats, and continued their study of the encephalitis, eventually demonstrating that it was caused by a virus now known as Murray Valley encephalitis virus, transmitted by mosquitoes, but in no way related to myxomatosis virus. The concurrence of the encephalitis with the spread of myxomatosis virus had been mere coincidence, an object lesson to alarmists. Both scientists gained considerable celebrity from this episode. But Burnett did even better when he later received the Nobel Prize in Physiology or Medicine for his work on the immune system.

What of the rabbits? By 1957, it was clear that Peter Rabbit had again bested Mr. McGregor. Rabbits were still dying of myxomatosis, but not nearly at the predicted rate, and not nearly fast enough to cause more than a modest reduction in their numbers. The number of wild rabbits in Australia soon rose to 300 million, at least ten on every acre of the continent, each one consuming as much grass as a sheep, all of them descended from those dozen released by Thomas Austin in 1859. And myxomatosis virus thrived with them. What caused this dramatic adjustment in the interaction between host and parasite?

All good Darwinians would expect that a relatively resistant strain of rabbit would emerge from this experiment in biological warfare, and that clearly happened. Something more subtle also occurred, however. By 1957, the myxomatosis virus isolated from wild rabbits in Australia had become less virulent. Mortality rates for infected rabbits dropped from 99 percent to 70 percent, a seemingly modest decline, but enough to assure that the virus would not eradicate itself by eradicating its host. Moreover, when tested in the newly emerged resistant strain of Australian rabbit, the virus caused mortality rates as low as 25 percent. In less than seven years, the virus and the host had evolved a dramatically different relationship.

In summary, two selections were operating in this interaction between host and parasite: selection for the more resistant host (as in survival of the fittest), and selection of the less virulent virus (so that the host might survive to perpetuate the virus). The host-parasite interactions we normally examine in humans have had much longer to reach equilibrium, but the features of each must reflect the pressures and accommodations that operated so rapidly in the early years of Australia's first encounter with myxomatosis.[45]

Meanwhile, Australians are again trying biological warfare against those long-suffering rabbits. This time they are using a distant cousin of poliovirus known as rabbit hemorrhagic disease virus, which is every bit as deadly to rabbits as myxomatosis virus and is also spread by mosquitoes. The first experiments with the virus were conducted on an island off the south coast of Australia. Every precaution was taken to contain the virus on the island. The results do not make good reading for biosafety committees. The virus got loose and began to spread like wildfire on the mainland—an estimated eight kilometers a day.

Not to be outdone by accident, the Australian authorities released the hemorrhagic disease virus throughout the continent, with stunning results. In some areas, the rabbit population has been reduced by as much as 95 percent. There has been no evidence of spread to other species, particularly Homo sapiens. But the population of feral cats has suffered a precipitous decline in many areas because rabbits had been their main prey. And complaints are being heard from Australian industries that were utilizing the wild rabbits to make pet food and hats.

Genes and Pathogenicity

The message from Australia is that the genetic programs of both host and parasite strongly influence the outcome of infection. This message addresses a wondrous gallery of mysteries. Why is it that fewer than 1 percent of the individuals infected with poliovirus develop neurological disease? Why is it that so many of us carry meningitis bacteria in our nasopharynx, but so few of us develop meningitis? Why is it that only one in every thousand children infected with measles virus develops encephalitis? Why is it that a certain small set of individuals infected with HIV never develop AIDS? The puzzles are as numerous as the microbial pathogens themselves.

We have a few glimpses of what the solutions to these puzzles may look like. Among the best examples are sickle-cell anemia and the thalassemias, diseases in which genetic defects have damaged hemoglobin—the molecule that carries oxygen in our red blood cells. But the same defects also make the cells relatively resistant to infection by the malaria microbe (a protozoan). As a result, natural selection has not eliminated the defects from the human gene pool. Instead, the diseases caused by the defects are prevalent throughout the populations that either reside or originated in geographical areas where malaria is endemic (in particular, equatorial Africa and the Mediterranean).

Similarly, the cystic fibrosis gene cripples a chemical pump in the intestine, but may also confer increased resistance to infection by the typhoid fever bacterium; and the gene responsible for Tay-Sachs disease creates a disorder in fat metabolism, but may also augment resistance to tuberculosis. Again, natural selection has preserved the aberrant genes despite their adverse effects. Resistance to microbial infections is apparently worth the price.

A less damaging means of resistance has been discovered for HIV. Some individuals have a seemingly innocuous genetic deficiency in the molecular gadgetry that allows HIV to enter lymph cells. These individuals are resistant to infection by the virus and the subsequent development of AIDS. It is worth noting that the concordant resistance to HIV and to AIDS offers potent evidence that HIV is indeed the cause of AIDS. As intimated before, a few scientists continue to argue that HIV may not be the cause of AIDS. This heterodoxy struck a sympathetic chord with some AIDS activists and has been espoused by the president of South Africa. But the evidence to the contrary is extensive and persuasive.

Parasites and Populations

Genes provide an intrinsic framework for the interaction between host and parasite. But there is another force that counts for at least as much, and that is the general welfare of the host. This fact belies the illusion that our burden of infectious disease has been reduced mainly by modern vaccines and therapeutics. Tuberculosis provides an archetype.

The frequency of tuberculosis declined drastically over the past century as nutrition, housing, and hygiene improved, and well before the introduction of antibiotics or vaccination. The disease lingered longest in populations of the poor. Now tuberculosis has reemerged with a vengeance, and the main victims are again the underserved—some of whom suffer from the additional liability of having immune systems compromised by infection with HIV. AIDS is a great leveler in this context, however, because even in the affluent, its effect on the immune system eventually outweighs any benefit that might result from a superior standard of living. Susceptibility to tuberculosis in the immunocompromised knows no class boundaries.

Standard of living plays a dominant role in all of human health. Over the past century, life expectancy in the United States has risen steadily to its current seventy-six years. Our increased longevity is due in large part to simple measures, all of which have improved our standard of living, and most of which have blunted the effects of infectious disease. In other words, we can relate the health of our populations as much to socioeconomic factors as to medical intervention. And where we have failed in our confrontation with pestilence, our failure has had more to do with standard of living than with a lapse in scientific ingenuity.

The disruptions wrought by microbes have repeatedly changed the course of human history. It was probably pestilence as much as any other single factor that accounted for the European conquest of the Western Hemisphere in the sixteenth century. Several factors account for the devastation that results when a host population encounters a virulent microbe for the first time. First, the absence of any prior immunity among the hosts allows the microbe to spread unfettered. Second, for reasons that are far from clear, infections often take a harsher toll in adults, and that toll will be especially grievous when none of the exposed adults possess immunity by virtue of infection in their youth. And third, a naive host population has not had the opportunity to evolve a gentler interaction with the microbe.

On the other hand, the Black Death may have fueled the burst of human creativity known as the Renaissance. At the time plague struck, medieval society had fallen into economic stasis, caused in large part by the "Malthusian deadlock" of dense population. The plague broke that deadlock by decimating the population, liberating land for diverse uses, creating the need for laborsaving devices, and unleashing the ingenuity of Renaissance society. The catastrophe of pestilence "gave to Europeans the chance to rebuild their society along much different lines . . . It assured that the Middle Ages would be the middle, not the final, phase in Western development."[46]

If pestilence is one of the great destabilizers of civilization, war is another. Yet even the disruptions of war are amplified by microbes. Florence Nightingale was among the first to appreciate and quantify this amplification. Nightingale is renowned for her image as a compassionate nurse—"the lady with the lamp." But she was also a pioneer in the scientific study of how infectious diseases alter human affairs.[47]

By rigorous analysis of mortality statistics, Nightingale demonstrated that pestilence inflicted more grief on the British Army during the Crimean war than did the wounds from battle. The same was true for the contemporaneous American Civil War, in which both measles and hepatitis caused far more deaths than gun powder and sabers, but for which there was no corresponding lady with a lamp.

The popular view of Florence Nightingale offers little hint of intellectual rigor or mathematics. But she was in fact highly numerate and the first woman to be elected to the London Statistical Society. Her work had a profound influence on the British military and has been called "the first example of someone using health care data to affect [sic] governmental reforms in the interest of preventing death and disease."[48] Lytton Strachey did her a great injustice with his demeaning portrayal in *Eminent Victorians*. In the words of a modern evaluation, Nightingale was "tough, canny, powerful, autonomous and heroic."[49]

Even our success in besting microbes can bring untoward consequences. Chief among these is a disturbance of population balance. For example, elimination of malaria from Mauritius led to a doubling of the population within a decade, even though the birthrate remained constant. Stated more broadly, relief from pestilence is a major factor in the population explosion that has threatened human welfare and for which no satisfactory remedy has yet been established. For the moment, the global epidemic of AIDS may provide a macabre counterbalance: the population of Africa faces decimation; and still emerging, but vast and largely uncharted epidemics of the disease are threatening India and China.

Foiling Microbes

Having identified the microbial enemy, humankind moved against it with four strategies: interdict the spread of pathogenic microbes;

immunize against specific pathogens; develop treatments for the instances where infection has already established disease; and attempt to eradicate various pathogens from human purview (an outcome that is rarely practicable—it has so far been achieved for only one pathogen, smallpox; and is imminent for only one other, poliovirus).

The spread of a microbe can often be prevented if the means of its transmission is known. The nineteenth-century physician John Snow was among the first to dramatize this truth.[50] Snow began his medical career as a general practitioner, but in time became a pioneer in both anesthesiology and epidemiology. He achieved renown first as an anesthesiologist, by administering chloroform to Queen Victoria during the birth of her children Prince Leopold and Princess Beatrice. He also experimented on himself with a variety of anesthetics, some of which induced euphoria, and many of which were highly toxic. But it was as a student of epidemics that Snow earned a permanent place in the medical pantheon.

During outbreaks of cholera in London, Snow meticulously traced the distribution of the disease house by house in affected neighborhoods, then compared these data to the sources of drinking water for the affected households. He found that one water company was delivering far more cholera than its competitors, and this company was alone in taking its water from the Thames River downstream of where the London sewage entered. Snow called this circumstance "The Great Experiment," with apparent sarcasm. He argued that contamination with sewage must be responsible for the outbreak of cholera, and his evidence is admired to this day.

It was the Broad Street pump, however, that allowed Snow to create a durable popular legend. He traced an outbreak of cholera in the Soho neighborhood to a communal well on Broad (now Broadwick) Street, into which raw sewage was seeping. According to tradition, Snow had the handle of the well's pump removed and the epidemic abated. This anecdote is among the most famous in the history of epidemiology and public health. Alas, historical fact is often more prosaic than legend. We now know that Snow may have had nothing to do with the end of the epidemic. The handle was apparently removed too late to have much of an effect: the epidemic was already in decline,

The Surgical Clinic of Professor Gross. In 1875, Thomas Eakins produced a large oil painting that shows the renowned surgeon Samuel David Gross at work (he is the dominant figure facing away from the action). Eakins rendered the scene with a realism that scandalized viewers of the time, but the painting is now regarded as a masterpiece and one of Eakins's best. The painting was contemporaneous with the careers of Koch, Lister, and Pasteur, and exemplifies the state of medical practice at the time these individuals and others began the modern assault on infectious disease: the surgeons worked in street clothes and with bare hands, seemingly oblivious to the threat of microbial contamination. The figure reproduces an India ink wash that Eakins executed in 1876, in order to have a copy of the painting. (Reproduced by permission of the Metropolitan Museum of Art, Roger Fund 1923, 23.94.)

as Snow himself reported. In the words of one modern epidemiologist: "Snow was riding to glory on the downhill slope of an epidemic curve."[51]

There can be no denying that Snow knew he was on to something: "[My findings] led me to the conclusion that [cholera] is propagated by the morbid poison which produces it being accidentally swallowed; that this morbid poison becomes multiplied and increased in quantity on the interior surface of the alimentary canal, and that it passes off in the ejections and dejections, to produce fresh cases of the disease in those who happen to take the morbid matter into the stomach."[52]

Snow also speculated with remarkable prescience on the nature of the cholera poison: "For the morbid matter of cholera having the property of reproducing its own kind, must necessarily have some sort of structure, most likely that of a cell."[53] It would be more than thirty years before Robert Koch would identify the "morbid matter of cholera" and show that it is indeed a reproducing cell, the bacterium *Vibrio cholerae.* The recalcitrant Max Pettenkofer was a contemporary of Snow's and should have paid him more heed.

Snow's accomplishment looms so large in the history of medicine that Londoners bestowed upon his memory their most loving honor —they named a pub for him, located on the square that once held the Broad Street Pump. The irony is that Snow was strictly abstemious.

Snow may have had the right idea about how to control cholera. But he was not much honored in his own time—his ideas were strenuously resisted by the medical establishment. And he apparently had little influence with the public authorities. Parliament abolished the London Board of Health because it was proving too aggressive in its advocacy of public hygiene—those water companies were not happy. The *London Times* approved: "We prefer to take our chance of cholera and the rest [rather] than be bullied into health."[54]

Vaccines and Microbes

Artificial immunization originated in the ancient custom of deliberately inoculating individuals with pus from the sores of smallpox. Since smallpox is formally known as variola (from the Latin *varius* for "spotted" or *varus* for "pimple"), the term "variolation" was coined for the inoculation. Variolation often induced a relatively mild disease, along with subsequent immunity to reinfection with smallpox.

Despite its prevalence in the Near East, inoculation against smallpox was little used in Europe until the intervention of Lady Mary Wortley Montagu early in the eighteenth century. While living in Turkey with her husband, the English ambassador, she learned of the practice, applied it to her children, and eventually brought it back to England. Her vigorous efforts to spread variolaton among the English (and perhaps the fact that she abandoned her husband) earned her many detractors, among them Alexander Pope and Horace Walpole. But in the eyes of history, she has emerged as an intelligent and enlightened woman, struggling for emancipation.

The enthusiasm of Lady Montagu could not hide the fact that, on occasion and in no predictable manner, smallpox inoculation / variolation backfired into lethal disease. The crude nature of the vaccine also resulted in the occasional transmission of syphilis and hepatitis. The potential for backfire caused justifiable public skepticism about the utility and safety of vaccination. But visionaries eventually carried the day. Some paid a steep price.

In 1756, the fire and brimstone preacher Jonathan Edwards was elected president of the newly formed College of New Jersey, later to become Princeton University. Eager to demonstrate his faith in modern science (an enlightened stance for any revivalist), Edwards volunteered to receive an experimental smallpox vaccine prepared from human lesions. He was inoculated, got smallpox, and died without ever taking office.

Forty years later, the faith that had failed Jonathan Edwards was revived by Edward Jenner, who had once studied with John Hunter. Cowpox is a disease of humans that resembles mild cases of smallpox and is generally acquired by milking cows. Jenner was aware of folklore that individuals who had suffered from cowpox appeared subsequently to be immune to smallpox. He tested this idea by using fluid from the cowpox of a milkmaid named Sara Nelmes (or in a less common rendering of the story, a cow named Blossom) to immunize a youngster by the name of James Phipps, then challenged young Phipps two months later with a hefty dose of virulent smallpox. Phipps proved immune, and the foundation for the eventual eradication of smallpox had been laid. Modern vaccination against smallpox has utilized not cowpox virus, but another relative of smallpox virus known as vaccinia, whose natural origins remain a mystery—the virus was first recognized as a distinct strain in 1939, after it had inadvertently replaced cowpox as the agent used worldwide for vaccination.

Jenner's effort to report his experience with James Phipps was rejected by the premier research journal of his time, *Philosophical Transactions* (as it would be now, both because a clinical study based on a single subject has no credibility, and because the experiment with young Phipps violated all modern standards for human experimentation). So Jenner successfully vaccinated several more children, including his eleven-month-old son, Robert, and then published at his own expense. The title of his paper employed a formal term for cowpox, *Variolae vaccinae* (inspired by the Latin term *vacca*, for cow), and it was from this that the terms vaccine and vaccinate were derived.

As a tribute to Jenner, Louis Pasteur later proposed that these terms cover immunization against all pathogens. The proposal was accepted and remains common usage today. Jenner and his contemporaries prepared vaccine by macerating material from cowpox lesions. We have yet to improve by much on this practice: one authority has described the modern smallpox vaccine as "dried calf pus." Prodded by the threat of bioterrorism, we may now do better. This is an urgent matter: smallpox has the potential to be an especially devastating biological weapon, if properly deployed.[55]

Jenner had no doubt about what he had achieved: "annihilation of the Small Pox, the most dreadful scourge of the human species, must be the final result of [vaccination]."[56] Thomas Jefferson agreed and wrote to Jenner in admiration: "Yours is the comfortable reflection that mankind can never forget that you have lived. Future nations will know by history only that the loathsome smallpox existed and by you has been extirpated."[57]

The confidence of Jenner and Jefferson has now been vindicated. The vaccination that Jenner pioneered has banished smallpox from the face of the globe. In October of 1977, a twenty-three-year-old Somalian named Ali Maow Maalin became the last recorded case of smallpox acquired by natural infection.[58] He survived. The eradication of smallpox represents a triumph for the World Health Organization, which organized the systematic vaccination of vulnerable populations around the globe. The cost has been trivial in comparison to the price that humankind might have paid if smallpox had been left at large.

Ostensibly, all that remains of smallpox now are two stocks, stored in Russia and the United States. There are suspicions, however, that samples of smallpox virus may be hidden elsewhere in the world, perhaps intended for nefarious purposes. The Russian and American stocks were originally slated for destruction in 1999, but the U.S. government engineered a controversial reprieve so that the virus would be available for further research, particularly in defense against bioterrorism.

Not everyone was pleased with the success of Edward Jenner. For example, in 1807, a London surgeon with the politically evocative name of John Birch published a tract entitled "Serious Reasons for Uniformly Opposing Vaccination." He summarized his argument as follows: "In the populous parts of the Metropolis, where the abundance of children exceeds the means of providing food and raiment for them, smallpox is considered as a merciful provision on the part of Providence to lessen the burthen of a poor man's family." George Bernard Shaw was not much better, calling vaccination "a particularly filthy piece of witchcraft."[59]

Fortunately, vaccination eventually gained the favor of most medical and public authorities. In due course, it was extended to many other ailments, becoming a major defense against infectious disease. In particular, vaccination promises eventually to eliminate the great pestilences of childhood, including poliomyelitis, German measles, measles, diphtheria, whooping cough, and bacterial meningitis. It was Louis Pasteur, more than any other individual, who built the intellectual bridge between the pioneering work of Edward Jenner and the triumphs of modern vaccinology.

Louis Pasteur and Vaccines

On July 4, 1885, a nine-year-old Alsatian child named Joseph Meister was bitten repeatedly on his hands, legs, and thighs by a rabid dog. He was rushed to Paris, where Louis Pasteur had been developing a vaccine for rabies. On July 7, Pasteur began treatment of Meister with the new vaccine, which employed spinal cord taken from rabbits dying of rabies and then dried for various periods of time. Drying the spinal cord inactivated the rabies virus. Thus, the shorter the period of desiccation, the more likely that the spinal cord still contained living virulent virus.

Pasteur began with infected spinal cord that had been drying for fifteen days, then administered progressively fresher material until, on July 16, he gave an inoculation of spinal cord that had been removed the day before from a rabid rabbit and must have contained abundant virulent virus. Pasteur claimed to have demonstrated the safety and efficacy of this protocol by tests in dogs. That claim was challenged recently after a review of Pasteur's laboratory notebooks.[60] The challenge raised doubts about the ethics of Pasteur's research and prompted spirited rebuttals from those who regard Pasteur as a paragon of science.[61]

However perilous the vaccination that he received, Joseph Meister exhibited no symptoms and returned to Alsace. He later became gatekeeper for the Pasteur Institute in Paris. In 1940, fifty-five years after the incident that gave him a lasting place in medical history, Meister was ordered by the German conquerors of Paris to open the crypt where the bodies of Pasteur and his wife are interred. Meister refused to do so and committed suicide instead.[62]

In developing his vaccines, Pasteur attempted to isolate "attenuated" variants that had lost their virulence but remained alive, in the hope that these could be used as effective immunogens. We still do the same today. The strategy is little different from that of Edward Jenner, who used a virus with naturally weak virulence (cowpox) to immunize against a highly virulent virus (smallpox).

Pasteur had some strange ideas about how attenuated vaccines might work, and some of his protocols for preparing these vaccines were flawed. But he correctly believed in the possibility of attenuated vaccines that would contain microbes with permanent, heritable changes in their properties; and in advocating this belief, he broke important new ground. In the end, Pasteur was the first to create a vaccine of any sort in the laboratory (an attenuated vaccine for chicken cholera) and the first to develop a vaccine in the laboratory for a human disease (the rabies vaccine, which Pasteur had hoped would be attenuated but knew was not; see later in this chapter). Pasteur also claimed to have developed a truly attenuated vaccine for rabies by passing the virus through monkeys, but that vaccine was never tested in human subjects.

The apparent success with Joseph Meister gained Pasteur worldwide acclaim and created an immense demand for his services. Individuals at risk of rabies flocked to Paris from all over the world to be immunized—nearly 2,500 came in the first twelve months after the Meister episode. Pasteur became and remains a national icon in France. He was treated like royalty for the remainder of his life; he lived to see the establishment of a new research institute that housed his work and still bears his name (at one point, he was said to be receiving more than 10 percent of all the funds spent on research by the French government); he received a state funeral of the sort normally reserved for kings and chiefs of state; and he and his wife were enshrined in the lavishly gilded crypt that Joseph Meister would not open for the Germans, located in the basement of the Pasteur Institute.

Given the nature of Pasteur's rabies vaccine, it is entirely possible that he may have killed a few individuals among the thousands whom he vaccinated during his career. A future premier of France, Georges Clemenceau, had Pasteur prosecuted on such a suspicion. The prosecution failed, but the suspicion lingers today.

Pasteur himself understood the risks he was taking. He had performed many experiments with his rabies vaccine in dogs and knew its lethal potential. While the immunization of Joseph Meister was in process, Pasteur's wife relayed his anxieties to their children: "My dear children, your father has had another bad night; he is dreading the last inoculations on the child. And yet there can be no drawing back now."[63]

Pasteur displayed further daring in his gamble that immunization against rabies would succeed even when administered subsequent to an exposure to the virulent virus. This tactic does not work for most infections, because the microbe does its damage before the immune system has had time to mount a response. But rabies has an unusually long incubation period—the time that elapses between infection and the appearance of disease. The long incubation period is due to the manner in which the rabies virus reaches the brain. The virus is deposited by a bite in the wound, enters nerve fibers, and makes a very slow progress up the fibers to the central nervous system—the transfer from skin to brain can take as long as a year.

Pasteur gambled that the prolonged incubation period would allow immunization to prevent disease, even if administered soon after infection rather than well before. He knew that he was working against the grain, but he had faith in his reasoning: "Owing to the long incubation, I believe that we will be able to render [infected] patients resistant with certainty before the disease becomes manifest."[64] The surmise on which Pasteur gambled proved correct and underlies the management of rabies infection to this day, which also utilizes immunization following exposure to the virus. It provides a salient example of Pasteur's biological insight.

In a remarkable twist of history, Pasteur used his rabies vaccine to immunize the man who later discovered the bacterium responsible for the bubonic plague, Alexandre Yersin. Pasteur immunized Yersin after he had cut himself while performing an autopsy on a woman who had died of rabies. Yersin survived, to achieve great success as a scientist. His name was eventually given to an entire genus of bacteria, including the species *Yersinia pestis*, the cause of plague and, perhaps, the Black Death.

It is notable that Pasteur relied almost entirely on animal hosts for his various experiments. He had no other recourse. The rabies vaccine itself was prepared from infected rabbits, a necessary practice that continued well into the second half of the twentieth century. (After many years of effort, scientists finally learned how to propagate rabies virus in test-tube preparations of human cells, and virus for the vaccine is now produced by this means.) The work of Pasteur and its immense influence on human health dramatize the importance of research on animals for medical progress. There would have been no

other way for Pasteur to ascertain the role of the rabies virus in causing disease, and without the use of animals, humankind would have
waited a century at least for the perfection of a rabies vaccine.

Regrettably, prophylaxis against infection often fails: because we are
indulgent and careless (witness the continued prevalence of venereal
disease); because there are many infectious agents against which we
still have no vaccine (witness the continuing absence of a vaccine
against HIV); and because of political and administrative failures
(many children, especially the poorest, are not vaccinated—we have
been remiss even in the United States). So therapeutic agents for infectious disease remain a vital part of the physician's armamentarium.
The development of those agents represents one of the great triumphs
of humankind and one of the great sagas in medical history.

Magic Bullets

In the spring of 1911, the composer and conductor Gustav Mahler was
taken from New York to Vienna to die. Mahler was so weak that books
had to be torn into individual pages so that he could hold them to
read. Physicians in New York had identified Mahler's ailment as a sys-
temic infection with the bacterium streptococcus, the microbial adversary of Semmelweis.

The same microbe had killed Mahler's older daughter four years before, engendering the sorrow and resignation that pervade his later
music (not that his earlier music is especially joyful). Mahler's own infection must have been his second with streptococcus. His heart already bore the scars of rheumatic fever, which we now know to be an
aftermath of streptococcal infection, and which prepared the stage for
Mahler's lethal second infection—the deadly microbes set up house-
keeping on the scars.

On the way to Vienna, Mahler stopped in Paris to consult a then renowned, but now mercifully forgotten, bacteriologist by the name of
Chantemesse. In a moment of exceptional insensitivity, Chantemesse
called Mahler's wife, Alma, to come and look through the microscope
at a sample of Mahler's blood. "Even I have never seen streptococcus
in such a marvelous state of development," Chantemesse exalted.[65]

Chantemesse's insensitive enthusiasm was duly recorded in Alma's diary but offered no consolation to any of the Mahlers. Gustav was dead within the month, at the age of fifty-one, his Tenth Symphony unfinished. At the time of Gustav Mahler's death, his ailment was incurable. Parisian physicians bombarded him with the recently discovered radium, to no avail. Now we could cure Mahler within weeks, with a single therapeutic agent—penicillin. We owe this dramatic contrast to a remarkable group of people.

Working at the turn of the twentieth century (and under the influence of Robert Koch, with whom he had trained), Paul Ehrlich conceived the strategy of the "magic bullet," a therapeutic aimed exclusively at the cause of a disease and harmless to normal tissues. The idea came to him when he realized that certain chemical dyes were staining bacteria preferentially in microscopic examinations of infected tissue. Ehrlich began to test dyes by the hundreds against various infections. His first success was to cure a single mouse that had been infected with the microbe that causes sleeping sickness. When tried in humans suffering from sleeping sickness, however, the same treatment did not work.

Then Ehrlich hit upon the idea of testing systematic chemical modifications of a single dye. He struck paydirt on the 606th permutation, which produced a drug that could cure syphilis and was eventually named Salvarsan. Because treatment with Salvarsan had sometimes lethal side effects, however, Ehrlich died a disillusioned man, regarding his life in science as a failure. The Nobel Prize he had received in 1908 for his pathbreaking work was apparently no consolation.

Gerhard Domagk picked up where Ehrlich had left off. He focused his attention on a single bacterium, streptococcus, inspired in part by the continuing prevalence of Ignaz Semmelweis's biological adversary, puerperal fever.[66] The first success came late in 1932, with the discovery of effective sulfonamides—the first commercially available antibacterial drugs, still widely used to treat infections of the urinary tract. The initial results with infected mice moved Domagk to rhapsody: "We stood there astounded at a whole new field of vision, as if we had suffered an electric shock."[67] It is noteworthy that Domagk was an employee of the Bayer pharmaceutical company, which had recruited him to work on antibacterial dyes, and had in general displayed far more vision about the future of antibacterial agents than the academic community of the time. (Bayer had previously given the world one of the most successful therapeutics of all time, aspirin.)

The Nazi government denied Domagk the opportunity to accept the Nobel Prize in 1939 (see Chapter 1)—indeed, the Nazis arrested and detained him for a week after he inquired about the possibility of attending the Nobel ceremonies. But Domagk received an award of a much greater sort when a precious experimental sample of the drug he had discovered was used to cure his daughter of a life-threatening infection with streptococcus—what could not be done for the patients of Semmelweis or for Gustav Mahler and his daughter.

It all began when daughter Hildegard, age seven, pricked her hand with a knitting needle. The wound became infected with streptococcus. The infection spread rapidly up the arm and then into the bloodstream. Swollen lymph nodes were lanced twelve times without effect, and eventually, the attending physicians urged that the child's arm be amputated. Domagk refused. Emboldened by the results of testing in mice, Domagk treated her instead with his newly discovered drug, Prontosil. Within two days, her fever had abated and she recovered without further incident.

Domagk had similarly dramatic successes with other infected individuals. But the medical fraternity proved skeptical. An early success with a patient at New York City Hospital was dismissed as coincidence, with the disparaging comment: "Isn't it fortunate that this happened here in the New York City Hospital, and not in some small hospital out in the sticks, where they really would have believed this German dye had made [the patient] better."[68] But as experience mounted, the efficacy of Domagk's discovery could not be denied. In December of 1936, Prontosil was used to cure Franklin D. Roosevelt Jr. of a severe streptococcal infection—a widely publicized episode that helped secure acceptance of Domagk's breakthrough. And in accord with Domagk's original motivation, an early clinical trial of Prontosil was conducted with postpartum women suffering from puerperal fever, with almost universal success. Medical science had produced its first therapy for the adversary of Ignaz Semmelweis.

The therapeutic agents pioneered by Ehrlich and Domagk originated as synthetic chemicals. The use of these agents eventually earned the sobriquet "chemotherapy." But even as Prontosil was carving its way into medical history, another strand of discovery was developing. The result would be "antibiotics," substances produced by microbes that can be used to treat infections.

We know that there is a brisk ecological competition among microbes in natural environments, so it comes as no surprise that evolution has armed microbes with weaponry against one other. These weapons were originally distinguished from other inhibitors of microbes as "antibiotics." The distinction between chemically synthesized drugs and antibiotics has blurred with time, however, because chemists now synthesize and modify the types of molecules first found in antibiotics. For physicians and the general public alike, antibiotic has become a generic term for any drug that can be used to inhibit microbial growth. The term chemotherapy, meanwhile, has been expropriated to describe the treatment of cancer with drugs.

Penicillin

It was Alexander Fleming who, in 1928, gave the world the first antibiotic derived from a living organism. Fleming was working under the aegis of an elderly majordomo of London physicians, Sir Almroth Wright. Wright was adamantly opposed to the idea of treating infectious diseases with drugs. He believed instead in therapeutic immunization—the use of vaccines to cure infectious diseases as opposed to preventing them. Therapeutic immunization has not stood the test of time well, although it remains one of the hopes in the battle against AIDS.

Wright did his best to discourage Fleming's interest in drug therapy. But having failed, he became Fleming's most ardent publicist in the years to come. The persona of Wright lives on in George Bernard Shaw's play *The Doctor's Dilemma*, where he is represented as the ludicrously self-assured physician who spends much of his time on stage repeating the injunction to "stimulate the phagocyte"—Shaw's sarcastic allusion to therapeutic immunization.

In early September of 1928, Fleming returned from a vacation and examined a petri dish that he had set aside before the holiday. The exact date of this historic moment is not known because Fleming never recorded it in his laboratory notebook (the most reliable effort at reconstruction places the event on September 3). What Fleming saw on the petri dish was a little pile of fungus producing a substance that inhibited the growth of adjacent bacteria known as staphylococci, very nasty pathogens for humans. In that moment of observation, Fleming set the stage for one of the most stunning advances in the history of medical science.[69] Every scientist lives in abiding hope of a similar moment.

Accident had created the opportunity, but Fleming was alert to its potential. To echo Pasteur, "Chance favors the prepared mind."[70] Fleming had an immediate intuition of what might be at hand. Like any other scientist with an eye on immortality, he first gave a name to his discovery: penicillin, derived from the name of the fungus, *Penicillium notatum*. He then preserved the petri dish with formalin, kept it through the ensuing years, and when fame came, donated the dish to the British Museum, where it still resides. Original reprints of the paper in which Fleming described his discovery now sell for thousands of dollars. In the spirit of his time, Fleming never thought to patent his astonishing discovery. Times have changed: modern biomedical scientists vigorously pursue patents on their discoveries, and they are regularly reminded to do so by their university employers.

Fleming eventually searched through countless fungi, seeking other inhibitors of bacterial growth. He never found one. Indeed, he never even found penicillin again. His discovery was an extraordinary stroke of luck that was to transform medical practice and human history. At first, Fleming's original isolate of the fungus was passed from laboratory to laboratory like precious ore. In time, however, scientists other than Fleming derived variants from the original material that produced much larger quantities of penicillin, and these greatly enhanced production of the antibiotic for clinical use.

Fleming recognized from the outset that penicillin might be useful, but he never pursued the possibility effectively. Instead, ten years later, it was a team at Oxford University that made therapy with penicillin a reality: Howard Florey, his wife Ethel Florey, Margaret Jennings, and Ernest Chain. Howard Florey was an experimental pathologist from Australia who was driven by the conviction that antibiotics could be used to treat infectious diseases. Ethel Florey was a practicing physician who helped supervise the clinical trials that eventually demonstrated the efficacy of penicillin. Jennings managed Florey's laboratory at Oxford University. Chain was the chemist without whom Florey could not have prepared penicillin of sufficient purity for clinical trials. The adjective sufficient is used loosely here: the preparations first tried on human subjects were not more than 2 percent penicillin by weight; the remainder was styled by Chain as "rubbish," which might well have been toxic, but was not.[71]

There was another, often forgotten hero in the story: Norman Heatley, an associate of Florey's whose ingenuity made it possible to produce usable quantities of penicillin under primitive conditions. Among his innovations, Heatley found that a common hospital bedpan made an ideal vessel for propagating the fungal source of penicillin. As production geared up, hundreds of metal bedpans were stacked floor to ceiling in the Florey lab. Faced with the need for even more containers, Florey commissioned a pottery to produce six hundred vessels modeled after bedpans. Heatley also developed the procedure that was used to extract active penicillin from the fungus, adapted dairy equipment to performing the extractions on a large scale, and devised a quantitative technique for the measurement of penicillin—a modification that is still used to assess the relative sensitivity of microbes to different antibiotics.

The culminating moment came in early 1941, when the penicillin produced at Oxford was used to arrest a life-threatening infection in a Cambridge policeman. It was to be a bittersweet moment: the first successful use of penicillin, but one that ended in tragedy. The policeman was forty-three-year-old Albert Alexander, who had been hospitalized for two months fighting a losing battle against a spreading infection.

The illness had started as a small sore at the corner of his mouth, which then became infected by staphylococci and streptococci. The microbes progressively invaded the tissues of Alexander's face, his eyes, and his scalp, necessitating removal of the left eye. Then the infection spread to Alexander's right shoulder and lungs. Now his life hung in the balance.

On February 12, 1941, therapy with penicillin was begun. Within twenty-four hours, Alexander was obviously better. His temperature dropped to normal and his appetite returned. By February 17, his right eye had recovered. It was clear that penicillin was effective. But by then, every scrap of penicillin produced by the Florey laboratory, the entire world's supply of the drug, had been exhausted. The last three days of Alexander's treatment had been maintained only by collecting all his urine and extracting the penicillin contained in it. Florey likened the situation to "trying to fill the bath with the plug out."[72] Then the supply of recycled penicillin was exhausted too.

For the next ten days Alexander's improved health continued, and there was hope that the five days of treatment had turned the tide. But then the lung infection returned and on March 15, 1941, Albert Alexander died of a widespread staphylococcal infection. Florey, Chain, and their colleagues stood outside the hospital room and wept. But the die was cast. Florey now knew that only logistical problems remained to be solved before penicillin could begin to save lives.

Florey felt a special urgency to produce penicillin in large quantities. Britain was mired in the darkest hours of the Second World War. In anticipation of a German invasion, Florey, Heatley, and Chain had made provision for smuggling the penicillin mold out of England by smearing samples of it into the linings of their coats. There was a desperate need for a new treatment of wound infections. But British industry was in no position to produce penicillin in bulk. So Florey traveled to the United States for help. He got it, in return for the patent rights that he too had failed to procure in advance. An abundance of penicillin was soon flowing from U.S. pharmaceutical factories.

The partnership between Howard Florey and Ernest Chain was one of the most historically important in all of medical science, but it eventually broke up, leaving Chain particularly bitter. The genesis of the rupture is not entirely clear, but there is no doubt that Chain felt slighted by Florey: Norman Heatley was chosen to accompany Florey on the trip to the United States, not Chain; and Florey rebuffed Chain's suggestion that they patent penicillin. Chain left Oxford shortly after the end of the war and spent the remainder of his life in Rome. He shared the Nobel Prize with Florey and Fleming in 1945, but even that seemed not to console him. The Floreys' marriage also went sour, although they stayed together long enough to consummate their historic work. In due course, however, the Floreys divorced, and, after a discrete interval, Margaret Jennings became Lady Florey (Florey had been given a life peerage and the Order of Merit).

Much has been made of the fact that neither Fleming nor Florey attempted to patent penicillin. But under British law of the time, it is unlikely that any patent application from Fleming would have succeeded. He had invented neither penicillin itself, a natural substance, nor any new procedure to produce penicillin. Nevertheless, he was considered saintly for never having tried for a patent. Florey and his colleagues, in contrast, would have had an ironclad claim for a patent. Florey chose not to apply because that would have violated the academic ethos of the time. When Chain appealed his case to other colleagues, he was accused of "money grubbing." Chain's desire for a patent arose in part from what he had learned of such matters from his father, an industrial chemist.[73]

It was Fleming who emerged with the greatest glory. Over the last decade of his life, he traveled the globe in triumphant progression, feted by universities, cities, and nations. Today, virtually every major European city has a street that carries his name. There is even a Fleming crater on the moon. Suitably, one of the early successes of penicillin was its use to save the life of a friend of Fleming, who had been moribund with meningitis. As treatment of the patient progressed, Fleming found it necessary to inject penicillin directly into the spinal fluid. The drug had never before been administered to a human by this means. So at Fleming's request, Florey tried it on an animal (the species of which is no longer known). Fleming did not wait for the outcome (just as well, since the creature died within hours), but instead proceeded with the injection and achieved one of the earliest "miracle cures" by penicillin. Fleming had obtained the penicillin by a direct appeal to Florey, who supplied the antibiotic and told Fleming how to use it. In the end, it was Fleming who had needed Florey—to propel him to lasting fame, and to render an exquisite personal service.

In the decades to come, further triumphs would follow. The discovery of streptomycin in 1943 produced a miraculous therapeutic for tuberculosis and earned Selman Waxman a Nobel Prize.[74] Pharmaceutical chemists became virtuosos at diversifying the structure of antibiotics in order to achieve activity against a broader spectrum of bacteria and to frustrate the mechanisms of antibiotic resistance. The medical armamentarium against bacteria and fungi now boasts a bewildering variety of chemical structures, nomenclatures, and applications. It remains far from perfect, but it would surpass even the wildest dreams of Paul Ehrlich and Gerhard Domagk.

Viruses proved the most difficult of microbial adversaries, largely because their intricate dependence upon normal cells made it difficult to repress their replication selectively. An early success was achieved with agents that could suppress life-threatening infections with herpesviruses. Then the devastation of AIDS added new impetus to research on antiviral agents, and success followed in remarkably short order, greatly extending the lifespan of most individuals infected with HIV. Now even the common cold faces mitigation by chemical antidotes—not a cure, mind you, but at least mitigation.

The Future

So we have well-forged tools for the isolation and characterization of microbial pathogens, for the interruption of their spread, for the prevention of disease caused by infection, and for the treatment of infectious diseases once they occur. But infectious diseases are still the third leading cause of death in the United States, and the leading cause worldwide. Infections are responsible for more than 12 million deaths annually in developing nations alone and represent 90 percent of the global disease burden. The global epidemic of AIDS has become one of the great plagues in the history of humankind. Venereal diseases are rampant, their effects only amplified by the advent of AIDS. A century after the work of Robert Koch, tuberculosis remains among the most common microbial causes of death (more people are now dying of tuberculosis than ever before in history), in part because the bacterium has developed a versatile resistance to antibiotics. Recurrent epidemics of influenza and diarrhea still kill by the tens and hundreds of thousands. Hepatitis viruses kill millions annually and cause liver cancer—among the most common of human malignancies. And the risk of food-borne infections has been rising steadily with the globalization of the food supply and the increased consumption of fresh produce.

Old microbial adversaries reappear unexpectedly—recall the 1994 outbreak of the Black Death in Surat. New adversaries present themselves with remarkable frequency: Legionnaires' disease, Lyme disease, toxic shock syndrome, Reye's syndrome, five new hepatitis viruses, AIDS, Kaposi's sarcoma virus, hantavirus pulmonary syndrome, *Helicobacter pylori* (the bacterial cause of both ulcers and cancer in the stomach), Ebola, West Nile virus, the transfer of mad cow disease from cattle to humans—none of these was known two generations ago. We can be sure that there will be more. There can be no end to pestilence in our lifetime, perhaps in any lifetime. As the ecological dynamics of our planet evolve, so do the vast hordes of microbes with which we share the planet.

Some of the new plagues are of our own making. The profligate and too often unjustified use of antibiotics both in the practice of medicine and in the rearing of livestock has resulted in drug-resistant adversaries that we are not likely to defeat in the near future. Many authorities consider the abuse of antibiotics to be one of the major plagues of our times, nothing less than a medical crisis. The dimensions of this plague are staggering: we put 70 percent of all the antibiotics produced in the United States into healthy livestock, creating a vast machinery for the selection of resistant organisms; fewer than 20 percent of U.S adults with sore throats can benefit from antibiotic treatment, yet almost 75 percent of these individuals are given prescriptions for antibiotics by their physicians; as many as 50 percent of all prescriptions for antibiotics in the United States are unnecessary; and more than 90 percent of pathogenic staphylococci are now resistant to penicillin and a variety of other antibiotics—there may even be strains of this deadly bacterium for which we presently have no effective drug.

The preeminent challenge is to replace the treatment of infectious disease with its prevention. We have failed in many instances to do this. Some of our failures have been in the realm of science. The earlier successes of vaccination have been followed by a series of ineffective efforts to master more subtle microbial pathogens, efforts that show no prospect of immediate success (the failures to produce vaccines against HIV and malaria offer telling examples). But the more damning failures are those that can be ascribed to socioeconomic inequities, derived in turn from a failure of political will. These are costly failures, because prevention is so much less expensive than treatment and the other wages of disease.

In 1967, the surgeon general of the United States declared the end of infectious diseases as a major threat to the public health and advocated shifting federal dollars to research on chronic diseases such as cancer. His optimism was clearly premature. But a shift did indeed occur, both in dollars and, perhaps more important, in attitude. The study of infectious disease and microbial pathogens lost much of its glamour. For several decades, biomedical scientists turned elsewhere in search of challenges.

Now change is in the air again. Scientists have mobilized to meet the global threat of AIDS, there is new vigor in the efforts to create a vaccine for malaria, and the threat of bioterrorism has dramatized the need for further research on microbes. The renewed assault on infectious disease will be greatly aided by remarkable recent advances: the decoding of both the human genome and the genomes from an ever-mounting number of microbes.[75] Together, these decodings will bring new insights into why our individual sensitivities to infection vary so widely, and new strategies for the attack on microbial pathogens— magic bullets that we could not previously have imagined. We will gain access to some of the deepest secrets of pestilence. The still unwon war against infectious disease will be fought in new ways.

CHAPTER 4

Opening the Black Box of Cancer

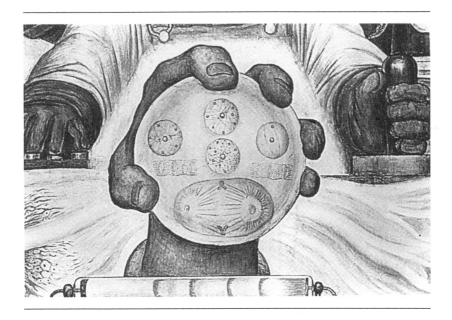

I propose to speak of a monster that is more insatiable than the guillotine; more destructive to life and health than the mightiest army that ever marched to battle; more terrifying than any scourge that ever threatened the existence of the human race. The monster of which I speak . . . has fed and feasted and fattened . . . on the flesh and blood and brains and bones of men and children in every land. The sighs and sobs and shrieks that it has exhorted from perishing humanity would, if they were tangible things, make a mountain. The tears that it has wrung from weeping women's eyes would make an ocean. The blood that it has shed would redden every wave that rolls on every sea. The name of this loathesome, deadly and insatiable monster is "cancer."

—Senator Matthew Neely of West Virginia, as quoted in
James T. Patterson, The Dread Disease

In 1966, Peyton Rous of the Rockefeller Institute received the Nobel Prize in Physiology or Medicine for his discovery of a virus that can cause cancer in chickens and animals. He opened his Nobel Lecture with the following words: "Tumors destroy man in a unique and appalling way, as flesh of his own flesh which has somehow been rendered proliferative, rampant, predatory and ungovernable. They are the most concrete and formidable of human maladies, yet despite more than 70 years of experimental study they remain the least understood . . . What can be the why for these happenings?"[1]

We now know the "why for these happenings." Over the past decade, a great change has occurred in how we think about cancer. Where once we viewed cancer as an unfathomed black box, now we have pried open the box and cast in the first dim light. Where once we thought of cancer as a bewildering variety of diseases with causes too numerous to count, now we are on the track of a single unifying explanation for how most or all cancers might arise. The track is paved with cells.

Cells

Robert Hooke first brought cells to public view.[2] Hooke was a seventeenth-century English physicist, meteorologist, engineer, architect, and biologist who also found time to fabricate some of the earliest microscopes. At a meeting of the Royal Society of London on April 15, 1663, Hooke placed a thin slice of cork under the lens of his homemade magnifier, revealing what he described as "little boxes or cells distinct from one another." The term "cell" caught on, and we still use it to describe the elementary building blocks of living tissues.

No single word now better embodies the sum of life: the image of a cell is as portentous as that of DNA. But there is an irony here. Hooke chose the word cell for its connotation of a rigid and static enclosure, which is what he saw in his microscope. Never has a connotation been less apt. Hooke was indeed looking at rigid enclosures, formed of the material we call cork. But the living units that had been within those enclosures, the immensely plastic units we now call cells, were gone. From the study of green plants, Hooke later came to appreciate that the "little boxes" contained a liquid of some sort, but he could not have realized that the liquid was part of a sophisticated machine for self-reproduction and the sustenance of life.

Hooke documented his many observations in a stunning book, *Micrographia or Some Physiological Descriptions of Minute Bodies, Made by Magnifying Glasses; with Observations and Inquiries Thereupon.* Samuel Pepys, a contemporary of Hooke and a renowned diarist, called *Micrographia* "the most ingenious book that I have ever read."[3] But most readers derided as trivial the immensely detailed drawings of fleas and other humble subjects that Hooke presented in

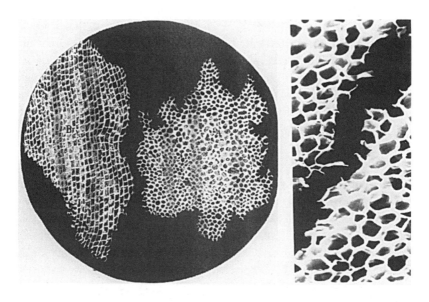

Cells. On the left, the interior of cork, with its empty "cells," as originally seen by Robert Hooke with a conventional microscope. On the right, the same scene, as viewed by a modern scientist using an electron microscope at low magnification.

his book. The microscope fell into disrepute as a scientific instrument, viewed as a plaything for the idle rich.

We have no reason to believe that Hooke ever appreciated the magnitude of his discovery of cells. Indeed, it would be two centuries before anyone else did. Then, in the period between 1835 and 1900, scientists at last perceived the nature of the cell and the central place of the cell in life on our planet. The perception came in several steps.

Matthias Schleiden, a student of plants, was perhaps the first to propose that cells represent the unit from which living organisms are built, an irreducible unit with a life all its own. Soon thereafter, Theodor Schwann adapted this proposal to animals. Still, the origin of cells remained a mystery. The popular view was that each cell formed de novo from the juices of the body, from a mystical substance known as the blastema in embryos, the cytoblastema in adults. But that view would not last. Its downfall was foreshadowed by the work of the great German embryologist Karl Ernst von Baer, who discovered the mammalian oocyte (egg).

Von Baer surmised that the growth and development of the embryo from a fertilized egg was dependent on continuous cleavage of some elemental component, whose nature he never grasped. But von Baer did realize that the search for those imagined cleavages would lead to what he called "the innermost tabernacle of embryology." Another German embryologist, Robert Remak, found his way into the tabernacle and eventually reached the conviction that all cells in the embryo arise from the division of existing cells. He published this conclusion in 1852 and changed the course of biology for all time.

The German pathologist Rudolf Virchow then generalized the scheme and gained much credit for it, perhaps because he publicized it so well: "A new cell can [never] build itself up out of any noncellular substance. Where a cell arises, there a cell must have previously existed *(omnis cellula e cellula)*, just as an animal can spring only from an animal, a plant only from a plant."[4] The Latin aphorism translates into "all cells come from cells," and it stuck. Good scientists are also often good marketers for their ideas.

The newly enunciated theory of the cell immediately illuminated several great problems in biology. Now it could be seen that all of life had a structural continuity. Every new individual arises from the union of two cells, the sperm and the egg. Cells are the living vessels of inheritance, whatever its chemical vehicle might be—a mystery that was to go unsolved for almost another century. And if all the cells of an organism arise from the single product of fertilization, then cells must possess an inherent ability to individualize themselves, to assume different functions, to "differentiate" (as biologists say); and they do this even though they share a common origin in fertilization and a common genetic dowry.

When I lecture to high school students, I find no aspect of biology more absorbing for them than the image of the sperm and the egg, uniting to engender a new individual. Skeptics might conclude that the student interest in fertilization is salacious. I concede that the mere mention of sperm brings a gleam to many an adolescent eye—the egg seems less provocative, perhaps because of a mistaken association with cholesterol. But I prefer to believe that students are moved by the wonder of embryogenesis. The fertilized egg, a single cell, multiplies myriad times over to produce the human organism: our form and function, good and bad; our consciousness, passions, intellect, and creativity; all rooted in the properties of individual cells—"The secrets of the mind are slumbering in the ganglion cell."[5]

Cells are also the material of evolution: they must change if species are to evolve. The genesis of that change is subtle because it arises from chance rather than experience. Our genetic dowry cannot be instructed by experience. We can force a zebra to stretch its neck for food, but the offspring of this zebra will be no more like a giraffe than was its parent. Lamarck was wrong.

It was August Weissman who gave cellular substance to these arguments when he embellished the theory of the cell with two lineages. One lineage assembles the body of each living creature and dies with that creature—the somatic lineage (from "soma," for body), a biological dead-end. The other lineage perpetuates the germ cells, sperm and egg, from one generation to the next—the germinal lineage or germ line, the carrier of our genetic dowry.

Weissman argued that the separation between these two lineages will last for all time. Changes in a somatic cell cannot be retrofitted to the inheritance of a germ cell. On this one point, he proved Charles

Darwin wrong, because Darwin had imagined that germ cells might collect information from somatic cells throughout the course of life, and that this information would then be part of the variation on which evolution relies. So far as we can tell, Weissman was correct, Darwin wrong (which hardly mattered, given the towering insight of evolution by natural selection that Darwin also conceived).

The formulation by Weissman brought to view for the first time an astonishing continuity that reaches from every cell now alive back into the depths of biological time, back to the primordial living matter from which we all arose. We have laid bare a kinship between ourselves and all of the other creatures that inhabit the earth, animals and plants alike; a kinship that was formed by the sculpting hand of evolution; a kinship that some would like to banish from our teaching because they believe it embarrasses their religious convictions.

The theory of the cell and the conclusions to which it led were based on inference, not experiment. The images that von Baer, Schleiden, Schwann, Remak, Virchow, and Weissman saw through their microscopes were static, frozen in time and space. Thus, the dynamic features of the cell theory as conceived in the nineteenth century were triumphs of the human imagination. A century more would pass before experimental science could give the dynamic behavior of cells tangible reality.

Cells and Cancer

Cells are the bricks with which all creatures are built—there are 300 trillion of these bricks in each of our bodies. But these are not ordinary bricks: they have an elaborate internal structure that allows them to live and breathe; they move from one place to another with purpose; they have distinctive personalities and assignments; they converse by means of chemical and molecular languages; and they multiply—ten thousand trillion times during the course of each human lifetime. The greatest wonder of cells, though, is that each knows what it is to do, and when and where. Cancer is a failure of the order that creates this wonder. The cancer cell violates its social contract with other cells, proliferating and spreading in an unfettered way.

The manner in which the proliferation and spread of cancer cells occurs was first appreciated in accurate detail by Wilhelm Waldeyer.[6] In 1867, Waldeyer published a microscopic description of how human breast cancer develops, beginning as a nidus of hyperproliferation in the glands of the breast, then proceeding to invade adjacent tissue, penetrate blood vessels, and spread to distant sites by transport of cancer cells through the lymphatic and blood vessels. Coming just a decade after the enunciation of the cell theory and produced with microscopes of dubious optics, Waldeyer's description was an astonishing achievement that has stood the test of time and could be little improved on today. But again, the images were static and the conclusions were inspired inference. To achieve a dynamic image of cancer cells and ascertain their individual properties, scientists turned to studies not in animals, but in petri dishes, applied in a way that Robert Koch could not have anticipated.

We can trace the beginning of this strategy to Alex Carrel, who in 1912 received the first Nobel Prize in Physiology or Medicine ever awarded to a scientist working in the United States. Although born and trained in France, Carrel was ostracized by the academic community there because of his pungent personality and his sympathies toward faith healing. He eventually settled at the Rockefeller Institute in New York City, faith healing no longer on his agenda. There he consolidated the pioneering work on vascular surgery and organ transplantation that earned him the Nobel Prize.

Carrel had learned the intricate stitching required for his work from the renowned lace makers of Lyon, one of whom was his mother. But Carrel left another legacy, one more pertinent to the study of cancer. He was among the first scientists to successfully propagate vertebrate cells outside of the living body, a procedure we now call "cell culture." He claimed to have kept one batch of chicken cells alive and propagating for thirty-two years. We no longer consider that claim credible, but there can be no denying the influence of his work. Carrel's personality never moderated. He spent his last years in France, espousing a toxic anti-Semitism.

Why has the cell culture pioneered by Carrel been so important to cancer research? The answer lies in simplification. Whole tumors are not easy objects for experimental study. So we resort to the assumption that the properties of individual cancer cells account for the behavior of tumors. We can define those properties by growing the cancer cells in glass or plastic vessels, using an artificial mixture of nutrients to feed the cells. Under these circumstances, cancer cells misbehave exactly as we might expect from the behavior of tumors in living organisms. The cells continue to grow even when crowded by their neighbors. They develop a very different appearance from their normal counterparts. And they behave like misfits, crawling over one another in a convincing caricature of the cells in an invasive cancer.

The early steps in the genesis of cancer probably occur in many of our cells during a lifetime, only to be aborted before matters get out of hand. But occasionally, the course of events continues to a lethal end, a homogeneous colony of cancer cells with the potential to expand unendingly. Biologists suspect that billions of cells may take the first step toward cancer in each of us during the course of our lives. Why then do any of us survive to tell the tale? The answer to that question has at least two parts.

First, one step is not enough. Several insults are required to produce a fully malignant cell, and the likelihood that these will combine in a single cell is very low. We will speak more of these combined insults later. Second, the immune system of our body can mount potent defenses against both foreign intruders (such as microbes and transplanted tissues) and errant natives (such as cancer cells). These factors combine to limit the frequency of cancer among humankind and to delay the emergence of most cancers until the later years of life.

What changes the cellular personality in a way that gives rise to cancer? Science has spent the last century trying to answer this question. Now, in a breathtaking sequence of discovery achieved over a brief period of time, an answer has emerged. All cancer can be attributed to a single underlying malady of the genetic program that directs the lives of our cells.

Genes and Cancer

The command center for our cells is located within a compartment known as the nucleus. Within the nucleus, the commands are carried on structures known as chromosomes. Individual human chromosomes can be released from the nucleus of the cell and stained so that each displays a distinctive pattern of bands that can be identified through a microscope—a sort of microscopic fingerprint. We Homo sapiens possess twenty-four different chromosomes, including the X and Y sex chromosomes, and each has a different fingerprint or "banding pattern." Normal cells possess two copies of each chromosome, with the exception of the sex chromosomes in males: male cells contain one X chromosome and one Y (female cells contain two X chromosomes).

Chromosomes carry the instructions that dictate the structure and activities of our cells. These instructions are inscribed on that remarkable molecule known as DNA. Several yards of DNA are crammed into each human cell; how the cramming is accomplished remains a great mystery. The instructions carried by DNA are composed of a chemical vocabulary that we call genes. The several yards of DNA in each of our cells harbor as many as 35,000 genes, which together constitute the human genome.[7]

The actions of genes are implemented by a universal molecular process that first copies DNA into smaller molecules known as RNA, and then RNA into even smaller molecules known as proteins. Proteins are the handmaidens of genes: the molecular expression of the genetic code, the molecules that get most of the jobs done, the components from which most of the cell is built, the engines that drive the chemical reactions of life.

It now appears that cancer results from mistakes in this chain of command. The mistakes originate from damage to DNA and the genes that it carries, damage that scientists call mutations. Usually, these are not mistakes that are inherited; instead, they occur at various times during our lives, gradually accumulating to a catastrophic threshold, beyond which lies the cancer cell. To recall the language of August Weissman, most cancers arise from genetic mistakes in our somatic cells rather than in our germ line. No more than 10 percent of all human cancers are inherited from one generation to the next.

How have these mistakes been found? Like so much else in science, initial progress came not from a deliberate search, but from unex-

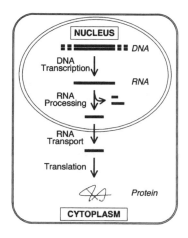

Expression of the genetic program in vertebrate cells. DNA in the nucleus of the cell is first copied ("transcribed") into the related molecule, RNA. The RNA is then "processed" into a smaller length by removing superfluous internal portions (known as "introns"). The processed RNA is transported to the cytoplasm of the cell and there copied ("translated") into the final effector molecule, protein. Each step in this sequence of events is executed by elaborate molecular machinery composed of dozens of different molecules. (Diagram courtesy of Ken Field.)

pected quarters. The story has five themes that converged to give us our newfound understanding of cancer. It is a story that is rich with the human face of science, and rich with lessons about how science is done.

The Cellular Inheritance of Cancer

The first and most fundamental of our five themes is this: the cellular property of cancerous growth is heritable. One cancer cell begets two others and so on ad infinitum. The allusion here to infinity is justified. Some human tumors have now been propagated in the laboratory through so many generations that the number of cells produced exceeds the number of stars in the known universe. Yet with rare exceptions, those cells continue as aggressively malignant as they were at the outset. There is no more dramatic example of constancy in all of the natural world. Paradoxically, cancer cells display inordinate plasticity as well, manifest as the ready appearance of resistance to therapeutics. We will explore later the mechanism of this plasticity and its role in the genesis of cancer.

Some historical sources credit Rudolf Virchow with the prescient suspicion that the behavior of cancer cells might be inherited. The suspicion originated from the fact that the cells of metastatic cancer, spread throughout the body, generally resemble one another as well as the cells in the original tumor when viewed through a microscope. Thus, all the cells of a metastatic cancer might be relatives: one cancer cell may have arisen from another: *omnis cellula e cellula*. This seems self-evident to us today. But in Virchow's time, it was beyond the pale.

Virchow has also been credited with the proposal that if all the cells of a tumor are relatives, then they must have originated from a single cell. This too would have been an eerie premonition of later observations. Most cancers begin as single wayward cells, whose astonishing fecundity engenders the billions of cells that finally compose a malignant tumor. Put into medical vocabulary, these tumors are "monoclonal," arising from single progenitors, rather than "polyclonal," arising from multiple cells that might have gone astray concurrently.

Despite these attributions to Virchow, his thinking in several areas was murky and calls into doubt his grasp of cellular reality. In particular, he never abandoned the idea that all tumors arise from one kind of tissue, even though they ultimately assume diverse identities (such as cancer of the breast, or lung, or colon); he coined the term "metaplasia" to describe the transition. We now know this to have been a fundamental error: most cancer cells have properties that actually reflect the tissue in which they originated, a testimony to their genetic lineage. Metaplasia has remained in the medical vocabulary to describe the transformation of one tissue into another (for example, the formation of bone within fibrous tissue in response to chronic irritation), but it is no longer soiled by Virchow's misconception.

Virchow's view of metastasis was also far off the mark. He refused to believe that cancer cells disseminate through the bloodstream as "embolisms" (a term he originated), arguing instead that cancer was spread by a fluid that could evoke malignant growth at remote sites in the body. It is difficult to see how this view could lead to *omnis cellula e cellula*.

Virchow was admittedly a very busy man and, thus, might be excused his confusions. He combined his studies of anatomical pathology with an interest in anthropology (he founded the anthropological societies of both Berlin and Germany); he invested great time and energy in promulgating his liberal political views, particularly in opposition to the authoritarian Bismarck; and he was the chief architect of the Berlin sewer system. But there was a limit to his enlightenment, and sadly, this was exemplified particularly by his outspoken opposition to Ignaz Semmelweiss in the matter of puerperal fever.[8]

Out of the germ of truth embedded in Virchow's ideas, there eventually emerged an overarching view of how cancer cells arise. The American pathologist Peter Nowell brought this view to its final form in the 1970s. The errant cell that initiates the genesis of a cancer is not in itself malignant. But the progeny of that cell continue to change, step by step. Each step represents a genetic event that is favored because it makes the cellular lineage more robust. Over time, the malignant state is reached: emerging cancer cells acquire the ability to proliferate indefinitely, to invade adjacent tissue, to disseminate as metastases, to evoke the provision of a new blood supply, to elude the defenses that usually eliminate incipient cancer cells. Thus, every tumor represents the outcome of an individual experiment in cellular evolution, driven by relentless selection for advantage.[9] "Nothing in biology makes sense except in the light of evolution," not even an outlaw like cancer.[10]

The Extrinsic Causes of Cancer

If cancer begins as a single cell that eventually progresses to a full-blown malignancy, then what drives the deadly sequence of events? At the outset, there seemed to be two possible answers: cancer might arise from spontaneous events intrinsic to our bodies—our cells might go astray of their own accord—or there might be extrinsic agents that elicit the mischief. There is no inherent reason that these two possibilities should be mutually exclusive, but for at least two centuries, the focus has been mainly on extrinsic causes.

In a magisterial summary published in 1981, Richard Doll and Richard Peto argued that 80 percent or more of cancers in the United States are in principle preventable because they arise from various extrinsic causes such as diet, lifestyle, personal habits, and environmental factors.[11] That view has gained great credibility in the interim. In addition, however, we have convinced ourselves that the various extrinsic causes of cancer might be united by a common mechanism; and paradoxically, that such a mechanism can also explain the competing intrinsic view of carcinogenesis.[12] The unifying mechanism is damage to DNA, the second of our converging themes.

In 1761, John Hill, a London physician, published the claim that inhalation of snuff caused nasal cancer. This is reputed to be the first formal report of an external cause of cancer. In our time, the evocative word "snuff" has been replaced by the disarming term "smokeless tobacco." But its consequences for human health remain every bit as ominous. Hill was anticipated in spirit by King James I of England, who in 1604 railed against the evils of smoking in an edict he entitled "Counterblast to Tobacco." The king also encouraged the public exhibition of human lungs blackened by tobacco smoke in an effort to discourage smoking.

Fourteen years after Hill's publication about snuff, Percival Pott achieved lasting fame when he reported that the chimney sweeps of Britain were highly prone to cancer of the scrotum, and attributed this to the soot of incompletely burned coal. French sweeps were said to be less afflicted, perhaps because they washed more frequently than the Brits (a national difference some say persists to this day). Danish chimney sweeps had a similar problem, which was ameliorated by the use of protective clothing in what is generally viewed as the first successful program for cancer prevention. Tobacco provided one additional hint when, in 1795, Samuel T. von Soemmerring noted that pipe smokers suffered an unusually high incidence of lip cancer.

These crude efforts at what we now call epidemiology were far ahead of their time, and recognition of their significance languished until late in the nineteenth century. Then the industrial revolution exposed humans to large quantities of noxious agents. Within decades, paraffin oils, mining dust, arsenic, aniline dyes, and asbestos came under suspicion. The importance of physical agents also became apparent, in the form of skin cancers attributed to excessive sunlight, as well as those affecting experimentalists working with the newly discovered X-rays. During the twentieth century, numerous physical and chemical agents were implicated as causes of cancer by observation of large populations. Two particular heroes of that story stand out from a meritorious crowd.

The first was Wilhelm C. Hueper, a German physician who immigrated to the United States in 1923 with a well-established interest in occupational disease.[13] Hueper was hired by the Du Pont company to pursue the relationship between exposure to chemicals known as aromatic amines and bladder cancer. But when his findings proved embarrassing to his corporate employers, he was fired. He used his enforced leisure to write a magnum opus entitled *Occupational Tumors and Allied Diseases*, the first authoritative treatise on the occupational causes of cancer.

Hueper spent the remainder of his career at the National Cancer Institute in Bethesda, Maryland, where his efforts to study occupational hazards were resisted and even censored. By the time he retired, however, his once heretical message about carcinogens in the workplace was national dogma, and concern about environmental carcinogens had become, if anything, overwrought. Hueper's story has a curious postscript: until virtually the end of his career, he vigorously discounted the mounting evidence that smoking is a major cause of lung cancer.

Hueper should have listened more carefully to our second hero, Ernst Wynder. While still a medical student in the 1940s, Wynder encountered a hint that there might be a connection between smoking and lung cancer (not a new idea even then, but one that had been paid little heed). He initiated his own study of the issue and quickly accumulated provocative data. Wynder solicited the patronage of one of his faculty—Evarts Graham, a distinguished thoracic surgeon and a heavy smoker. Graham was at first skeptical, but when he saw the final results from Wynder's study, he bought the argument and stopped smoking (to no avail—Graham died of lung cancer a few years later).

Wynder and Graham published their first set of data in 1950, claiming that smoking could increase the risk of lung cancer by as much as fortyfold. Within six months, similar results were announced from England, and the case against smoking has grown steadily stronger ever since. Wynder devoted his long career to the relentless pursuit of tobacco as a carcinogen, and he paid dearly for that pursuit. He was dismissed by the medical community, maligned by the tobacco industry, and harassed by the director of the research institute where he worked (the Sloan-Kettering Institute in New York City, which was receiving generous donations from tobacco companies). But Wynder's views would prevail. He lived to see the surgeon general of the United States issue an official advisory that "smoking is causally related to lung cancer" and the gradual acceptance of that view by government authorities and the public. The tobacco industry survives him, energetically fighting to reverse a decline of sales in the United States and to expand its market in developing nations.

Experimental Carcinogenesis

Through occupational and recreational exposures, humankind had inadvertently used itself to perform numerous experiments on carcinogenesis. But could the implications of these experiments be authenticated in the laboratory? The question was answered first by the French scientist Jean Clunet, who in 1908 reported that deliberate exposure to X-rays could induce skin cancer in rats. The initial experiment was a modest one, to say the least. Clunet bombarded four white rats with X-rays. Two of the rats survived, and one of these developed a tumor at the site of irradiation. The work would not pass muster in our more rigorous times. But it was published and soon authenticated by replication.

Experimental carcinogenesis with chemicals applied to the skin proved more difficult, in part because early trials were performed with rats and dogs, species we now know to be exceptionally resistant to the chemicals. In 1915, however, Katsusaburo Yamagiwa and Koichi Ichikawa in Japan reported the induction of cancer in the skin of rabbit ears by the application of coal tar. They had fortuitously chosen a sensitive species and a vulnerable anatomical site, and they had been persistent: success came only after application of the carcinogen every two or three days for a period of more than one hundred days. Soon thereafter, a student of Yamagiwa demonstrated chemical carcinogenesis in the skin of mice, a more tractable system than rabbit

"Cancer was produced. Proudly I walk a few steps." Haiku by Katsusaburo Yamagiwa, 1915. Illustration courtesy of James Miller.

ears. The exploration of chemical carcinogenesis was now properly launched.

Elated by these discoveries, Yamagiwa wrote a haiku in his own masterful calligraphy. It read: "Cancer was produced. Proudly I walk a few steps."[14] Yamagiwa's elation was justified. His findings are regarded as a landmark in the history of cancer research because they pointed the way to the identification and characterization of chemical carcinogens, and because they exemplified how multiple, infrequent events are required in the genesis of cancer (see later).

In the two generations spanned by Virchow and Yamagiwa, medical research had demystified the fundamental nature of cancer and progressed to the first experimental induction of the disease by Clunet and Yamagiwa. But sometimes there is no justice in contemporary judgment. In 1913, the Danish scientist Johannes Fibiger reported that he could induce stomach cancer by feeding worms to rats. The work is now regarded as an embarrassment. But it was Fibiger who received the Nobel Prize for the development of experimental carcinogenesis in 1926, not Clunet or Yamagiwa.

The pioneering work of Yamagiwa was performed with crude coal tar and mixtures of organic chemicals. No one knew the exact nature of the offending agent. But in 1930, Ernest Kennaway and his colleagues in London purified a carcinogen from coal tar and identified it as the organic chemical dibenzanthracene. They then synthesized the same chemical and demonstrated that the synthetic material was also carcinogenic in rodents. Here at last was rigorous identification of an individual chemical carcinogen. Soon thereafter, carcinogenesis affecting internal organs (particularly the liver and bladder) was achieved by feeding chemicals to rats. The generality of carcinogenesis by external agents was now well established.

As with Louis Pasteur, here again we see illustrated the vital role of experiments with animals in the progress of medical science. It is fair to say that without such experiments, we would still know very little about the identity of the carcinogens in our daily existence, and we would still be searching for an explanation of why these agents cause cancer.

Carcinogenesis and DNA

The discovery of external carcinogenesis immediately raised the issue of how the carcinogens might be acting. Again, pride of place belongs to X-rays, whose ability to damage genes was demonstrated by Hermann Joseph Muller in 1928 using *Drosophila melanogaster*—the fruit fly. Muller was the first person to deliberately induce mutations in genes by any means, and the ability of X-rays to do this prefigured the eventual discovery of how they cause cancer. At the time, Muller had little scientific interest in cancer and the creature with which he was experimenting had never been used for cancer research. Yet his discovery underlies all our current thinking about how cancer arises. He had shown for the first time that a carcinogen (X-rays) is also a mutagen.

In contrast to Yamagiwa, Muller took little comfort from his discovery. There was to be no triumphant haiku for him. Instead, at the age of forty-one, awash with fame but despondent over competition with his former research mentor (Thomas Hunt Morgan of Columbia University), criticisms of his work, a failing marriage, and the political turmoil of the Great Depression (Muller was an ardent socialist and abhored the political milieu of Texas, where he was then working), Muller took an overdose of sleeping pills and wandered off into the woods outside of Austin. He carried a suicide note saying "my period of usefulness, if I had one, now seems about over."[15] His colleagues organized a search posse (this was Texas, after all) and found him the next day, dazed but still alive. Fifteen years later (1946), Muller received the Nobel Prize for his discovery of experimental mutagenesis. This time, Stockholm got it right.

Within a decade of Muller's discovery, the chemical carcinogen 3-methylcholanthrene had also been shown to be mutagenic (this time, in mice). And by 1948, prominent geneticists were asserting that just as all carcinogens are probably mutagens, all mutagens would likely prove to be carcinogenic. Muller had actually made the same suggestion in his original publication on X-rays—he was not oblivious to the biological puzzles posed by cancer; he simply was not originally motivated by them. Chance does favor the prepared mind.

Then the work hit a snag. Why do chemical carcinogens cause mutations? The answer was obscured by two difficulties. First, some of the chemical carcinogens were inert in the laboratory—they entered into chemical reactions only with great difficulty. Why then did they have biological effects? And second, what was the molecular target of the carcinogens? Would it be the same as the target for mutagens? Could mutagenicity and carcinogenicity be equated?

The first of these puzzles, the chemical puzzle, was solved with the discovery that the body often betrays itself. In the process of detoxifying chemicals, it can create highly reactive intermediates that represent the actual carcinogens. The possibility of this metabolic activation was first suggested in 1935 by the English scientist E. Boyland, then put on a firm experimental ground three decades later, particularly by Elizabeth and James Miller in Madison, Wisconsin.

Identification of the reactive forms of chemical carcinogens helped clarify the search for their molecular targets, but progress was still slow. The idea that proteins were the crucial targets for carcinogens arose and proved to have great staying power. It yielded the field to genetics only slowly, as the evidence for the mutagenicity of carcinogens continued to mount. Any poll of modern biologists would produce a rousing majority for the view that carcinogens act by affecting DNA. But still, there are lingering difficulties with that view.

No one anticipated the difficulties when, beginning in 1975, Bruce Ames and others developed tests that allowed the detection of mutagenicity in petri dishes rather than animals. The stage seemed set for a definitive test of the relationship between mutagenicity and carcinogenicity. And at first, there appeared to be a direct correlation between mutagenicity in these simple tests and carcinogenicity in rodent models. Simply put, the more likely a substance was to damage DNA, the more likely it was to cause cancer. Or so it seemed. Alas, matters are no longer so clear. We now know that perhaps half of all the substances that are carcinogenic in rodent tests do not score as mutagens in an "Ames test," a discrepancy that remains inadequately explained (although there are splendid hypotheses on record).

These lingering ambiguities have caused no end of difficulties in the efforts to identify carcinogenic agents and regulate their use. They also illustrate why efforts to study the mechanisms of external carcinogenesis have contributed only tangentially to the search for the inner malady of cancer cells. Indeed, if the evidence were to end here, we would still be swimming in doubt.

But the day was saved when another line of inquiry emerged to provide more substantive clues that the genetic apparatus is at fault in tumorigenesis. This was the microscopic study of chromosomes, "cytogenetics," which eventually produced the observation that cancer cells frequently harbor abnormal chromosomes. We have arrived at the third of our converging themes.

Chromosomes and Cancer

In 1903, Walter Sutton published a paper entitled "The Chromosomes in Heredity."[16] By studying the chromosomes of grasshoppers, Sutton had reached a series of landmark conclusions now taught in every high-school biology course: (1) with the exception of sperm and egg, all cells of metazoan organisms contain two sets of chromosomes; (2) each cognate pair of chromosomes is physically distinctive; (3) when cells divide, each daughter cell gets a complete complement of chromosomes, one member of each pair derived from the mother, the other member from the father; (4) each sperm or egg receives only one set of chromosomes, but the number is restored to two at fertilization; and (5) all of these facts combine to suggest chromosomes as the likely carriers of inherited traits—a truly pathbreaking conclusion.

At one stroke, Sutton had created the science of cytogenetics and identified the genetic apparatus of our cells. At the time of his momentous publication, Sutton was a twenty-five-year-old student at Columbia University. He conceived, conducted, and published his work alone. But having transformed our understanding of the genetic apparatus, he never published another paper. Instead, he went on to become not a scientist, but a surgeon; and to die prematurely at the age of thirty-nine. He left behind insights that can easily stand in the history of genetics with those of Mendel, and of Watson and Crick.[17]

We can credit Sutton with the creation of cytogenetics, but it was Theodor Boveri who took that pursuit into the realm of cancer. In the same year as Sutton's publication, Boveri produced an intuition that still reverberates through the world of cancer research. Boveri had set out to study fertilization and the division of cells, using ascaris worms and sea urchins. His strategy was to force the fertilization of eggs by two sperm rather than one. This caused the fertilized cell to divide into four rather than two. As a result, none of the cells received a proper complement of chromosomes and cell division faltered.

Like Sutton, Boveri reached the important conclusion that chromosomes might be the carriers of individual genetic traits. More to our point, however, Boveri concluded his first report with the inspired speculation that cancer might be due to abnormalities of chromosomes, particularly an excess or deficiency of chromosomes in individual cells. No one quite knows where the idea came from, because Boveri was not studying the chromosomes of cancer cells. But it proved to be remarkably prescient.

To Boveri's distress, his idea drew little or no comment from his contemporaries and prompted no experiment. So in 1914, Boveri expanded his speculation into one of the most famous books in the history of biomedical science, entitled *The Origin of Malignant Tumors*; in the canon of cancer research, it is easily equivalent to the *Principia* of Newton in classical physics. The central argument was as follows:

> The unlimited tendency to rapid proliferation in malignant tumor cells [could result] from a permanent predominance of the chromosomes that promote division . . . Another possibility [to explain cancer] is the presence of definite chromosomes which inhibit division . . . Cells of tumors with unlimited growth would arise if those "inhibiting chromosomes" were eliminated . . . [Since] each kind of chromosome is represented twice in the normal cell, the depression of only one of these two might pass unnoticed.[18]

To modern students of genetics and cancer, these are breathtaking conclusions: in the assumption that specific chromosomal elements govern cell division; in the clear description of what we now call genetic dominance and recessiveness, decades before these concepts became clear from experiments in fruit flies; and in the anticipation by almost a century of the genetic malady in cancer cells.

The prescience of Boveri finally became clear in 1960, when Peter Nowell and David Hungerford teamed up to identify the "Philadelphia Chromosome," an abnormality found consistently in cells of chronic myelogenous leukemia, and the first chromosomal anomaly specifically associated with any neoplasm, named for the city in which it was discovered. The optical resolution available to Hungerford and Nowell failed to reveal the detailed nature of the Philadephia Chromosome, so it was 1973 before Janet Rowley showed that the abnormality arises from a physical mishap known as reciprocal translocation. Chromosomes 9 and 22 exchange portions of one of their arms, and one product of this exchange is the Philadelphia Chromosome. Why and how this happens, we do not yet know.

We do now know, however, of more than two hundred chromosomal aberrations that are consistently associated with one or another type of cancer. Each of these represents an explicit manifestation of the genetic mayhem in cancer cells. In many instances, the physical joints formed by the translocations and the genes residing there have been isolated by using the procedures of recombinant DNA. In making these isolations, scientists are drilling to the very core of tumori-

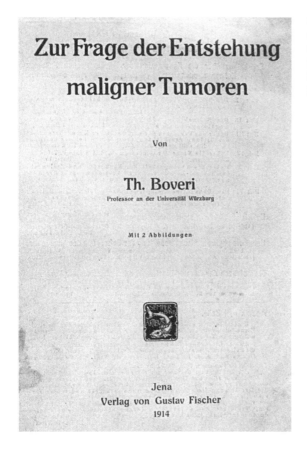

Chromosomes and cancer. Title page of the monograph that gave Theodor Boveri a lasting place in medical history.

genesis, spotlighting genes that might be contributing to the malady. These are extraordinary developments. When I graduated from medical school in 1962, cytogenetics was not yet even a diagnostic procedure, and the possibility of isolating pieces of DNA from focal points of a chromosome was beyond even fantasy.

Cytogenetics succeeded in pointing to the genetic apparatus as the ailing organ of the cancer cell. But the implication that malfunctioning genes might propel neoplastic growth could not be tested with microscopy. Instead, it was the study of viruses that produced the first explicit example of cancer genes—the fourth of our converging themes.

Viruses and Cancer

At the turn of this century, the "germ theory" of disease had gained great currency, owing mainly to the work of Robert Koch in Germany and Louis Pasteur in France. It seemed for a while as if all diseases might have microbial causes. Since no effort to implicate bacteria in cancer had succeeded, it was only natural that the discovery of viruses in 1898–1900 would lead immediately to inquiries about whether these agents might cause cancer. The first step was to ask whether cancer could be transmitted from one host to another by tumor extracts that had been filtered to remove bacteria and other cells. The only microbes that were likely to pass through the filters were viruses.

The first success came in 1908, when the Danish scientists Vilhelm Ellerman and Oluf Bang reported that they could transmit leukemia from one chicken to another with an infectious extract of blood cells. The work was dismissed as unimportant because leukemias were not in those days considered to be malignancies, and because chickens were not interesting. Peyton Rous at the Rockefeller Institute in New York City thought otherwise. In 1909, a farmer from Long Island appeared at the institute with a prize Plymouth Rock hen that had developed a tumor in the muscle of its right breast. The farmer hoped that the hen might be cured. He was referred to Peyton Rous, one of the few scientists at the Institute who had displayed an interest in cancer. After what must have been some slick talking, Rous killed the chicken and performed two landmark experiments.

First, he succeeded in transplanting the tumor cells from one chicken to another. This was a notable achievement, particularly because Rous had the good sense to perform the transplantations with chickens from the flock of the same farmer, thus displaying both an early awareness of transplantation immunity and negotiating skills of the highest order.

The second experiment took Rous to a new level of immediate controversy and eventual triumph. He prepared an extract of the tumor cells and passed it through a filter, much as Ellerman and Bang had done with the chicken leukemias. Rous then injected the extract into healthy chickens, which proceeded to develop tumors similar to those in the original hen. It appeared that the filtered extracts contained an infectious agent—more than likely, a virus—that could elicit tumors. The experiment succeeded only after the tumor cells had been transplanted several times from one chicken to another, suggesting that the infectious agent might have emerged during the transplantations. It would be three decades before the deep significance of this nuance became apparent from the work that earned the Nobel Prize for Harold Varmus and myself—more of that shortly.

Rous realized that he had discovered a virus that causes cancer.[19] Neither Rous nor anyone else of his time really knew what a virus might be like. For them, it was merely an invisible poison (the meaning of "virus" in Latin), a poison that seemed to have a life of its own, and in Rous's case, a poison that could cause cancer—the first cancer virus brought to proper view. From his discovery, Rous constructed the argument that perhaps viruses cause cancer in humans as well. The scientific community of the time was dismissive. They viewed the findings in chickens as irrelevant, the suggestion of human cancer viruses as nonsense.

Henry James proved more perceptive. The expatriate American author toured the Rockefeller Institute in 1910 and was introduced to Peyton Rous at a time when the youthful Rous was in the midst of his work on the chicken tumor virus, whereas James was deep into the miseries of age and but two years from his death. When James was told that Rous was in charge of cancer research at the Rockefeller, he responded fervently: "How magnificent! To be young and to have divine power."[20]

The microbe discovered by Peyton Rous is the archetype for a family of viruses known as "retroviruses." The viruses are named and renowned for their ability to reverse the flow of genetic information by means of the enzyme "reverse transcriptase," which is encoded by a retroviral gene and contained within the virus particle. The discovery of reverse transcriptase in 1970 startled biomedical scientists because it upended the "central dogma" of molecular biology, according to which the flow of genetic information was held to be undirectional, from DNA to RNA.[21] Reverse transcriptase reverses this flow by copying RNA into DNA, and in so doing, facilitates the replication of the genes of retroviruses, which are carried in these viruses as RNA rather than DNA.

Rous eventually abandoned the study of the virus he had discovered, which we now call Rous sarcoma virus, and pursued other forms of cancer research. It is often said that he gave up the study of his chicken virus because of ridicule. But according to his Nobel Lecture, Rous abandoned the work because he could not detect viruses in cancers of rodents.[22] He felt that if the phenomenon he had discovered was not universal, it was not worthy of pursuit.

Rous was perhaps right in principle, but he was wrong in reality. Over the next seven decades, a hard-fought intellectual battle slowly authenticated the ability of diverse viruses to cause cancer. The susceptible species include Homo sapiens, as Peyton Rous had once postulated. The battle had to overcome deeply entrenched preconceptions. For example, in the 1940s, John Bittner deliberately disguised his discovery of a virus that causes breast cancer in mice by calling it "milk factor" in all of his publications. Asked to explain this decades later, he remarked: "If I had called it a virus, my grant applications would have automatically been put in the category of 'unrespectable proposals.' As long as I used the word factor, it was respectable genetics."[23] Bittner was also eager not to offend his supervisor, Clarence Cook Little, who thought that the idea of cancer viruses was nonsense, and who controlled Bittner's budget.

Among those who kept alive the idea of cancer viruses was Ludwik Gross, a Polish refugee from the Second World War who found his way to the United States in 1940 and brought with him a zealous commitment to the viral cause of cancer. At the time, the fortunes of tumor virology were at a low ebb. Posted by the U.S. Army to the Bronx Veterans Hospital (an improbable venue for pathbreaking research), Gross began a series of clandestine experiments in which he attempted to transmit mouse leukemia by filtered extracts of leukemic cells.

After repeated failures, Gross succeeded by using newborn mice, whose immunological immaturity increases their susceptibility to induction of leukemia by retroviruses, and by fortuitously using a strain of mouse that is genetically vulnerable to infection with the leukemia virus that Gross had finally uncovered. The discovery was greeted with disdain and disbelief that dissipated only after the work was repeated by a more established scientist, Jacob Furth, who took the trouble to replicate Gross's protocols exactly and to use the same strain of mice. The authentication of Gross's discovery fueled numerous efforts to detect retroviruses that cause leukemia in humans, virtually none of them successful to date. (The single exception is human T-cell leukemia virus, which causes a rare malignancy of lymphatic cells.)

Peyton Rous was eventually honored with a Nobel Prize at the age of eighty-five (his son-in-law, Alan Hodgkin, beat him to the punch by three years), just as I was beginning my own career in research. (I remember seeing an announcement of the prize on a university bulletin board and realizing that I knew nothing of this man or his work. It was not long before I had rectified this shortcoming.) But Rous was also memorialized in a way that I find more notable. He has been enshrined in literature, portrayed as the character Rippleton Holabird in *Arrowsmith*, Sinclair Lewis's romantic novel about life in science. Although Rippleton Holabird has a large part in the plot, he is not a very attractive character. We apparently owe that depiction to Paul de Kruiff, the microbiologist who coached Sinclair Lewis during the writing of *Arrowsmith*. They put virtually the entire staff of the Rockefeller Institute into the novel, none of them in a good light. De Kruiff had once been on the staff at the Rockefeller Institute but had not fared well.

For good measure, de Kruiff left us his own analysis of why Rous gave up research on cancer viruses. "Dr. Rous was so amazed at his own discovery that it was rumored he couldn't stand the mental strain of going on with it and you couldn't blame him."[24] So it seems that de Kruiff too recognized how aversion to risk can impede scientific discovery (I have written of this in Chapter 2). It seems unlikely, however, that risk was much of a deterrent to Peyton Rous, who was reputed to be an immensely self-assured, sometimes arrogant individual.[25] To my great regret, I never met the man.

Enemies Within

As the reality of tumor viruses became secure, two schools of thought sprang up. One school argued that we should search for viruses in human cancer, that viruses must be a common cause of this disease. The other school held that since there appear to be many causes of cancer, we would be better off to use viruses in search of the central molecular mechanisms by which the disease arises. Against all reasonable odds, both views have been vindicated. On the one hand, viruses of many sorts—and most not retroviruses—have been implicated in a number of human cancers. Cancers of the liver and the uterine cervix are the most prevalent examples at present. On the other hand, the experimental analysis of tumor viruses has led us to the heart of the cancer cell, helping us to ferret out the molecular anomalies that engender its life-threatening behavior.

The utility of viruses for the experimental study of cancer arises from genetic simplicity, much as was described for poliovirus in Chapter 3. The DNA of human cells contains more than thirty thousand genes. Each gene has its own specific chore, and among these chores, there must be many that are important in the genesis of cancer. By contrast, viruses generally have fewer than a dozen genes, and only a subset of these genes is usually required to produce cancer. So viruses can simplify the search for genes involved in cancer by more than a thousandfold.

The power of this simplification became apparent in 1970, when Steven Martin demonstrated that Rous sarcoma virus possesses a gene that is responsible for both the initiation and maintenance of cancerous growth, but not for the proliferation of the virus itself. He did this by isolating mutant strains of the virus that could convert cells to cancerous growth at one temperature (35 degrees centigrade), but not at another (42 degrees centigrade). Moreover, even if cancerous growth was first established at the lower temperature, it could be reversed at any time thereafter by switching the cells to the higher temperature.

These findings signal the presence in the virus of a gene whose protein product is responsible for both the induction and preservation of cancerous growth. A mutation in the gene has rendered the protein

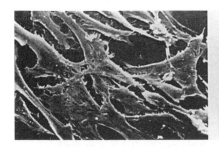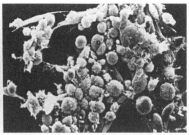

Normal and malignant cells as viewed with an electron microscope. The cells have been propagated on a flat surface. Those on the left are normal; those on the right have been infected with Rous sarcoma virus and converted to a cancerous form. Image magnified approximately 1,000-fold. (Electron micrograph reproduced by permission of Steven Martin.)

sensitive to the higher temperature. Here then was a true cancer gene, an "oncogene" in contemporary parlance. That gene was dubbed *SRC* (pronounced "sark") because it causes tumors known as sarcomas. Martin did this work against the advice of his postdoctoral mentor. We can only rejoice that the mentor did not pay too much attention to what Martin was actually doing.

It soon became apparent that the Rous sarcoma virus contains only four genes. Three of these are used to reproduce the virus, the fourth is *SRC*. Then scientists produced a map of how the four genes are arrayed along the RNA genome of the virus. Suddenly, a cancer gene had become a tangible reality. The discovery that the virus of Peyton Rous uses a gene to elicit cancer brought clarity to what had been a muddled business. There had been hints before that the elemental secrets of cancer might lie hidden in the genetic dowry of cells. But here in Rous sarcoma virus was an explicit example of a gene that can switch a cell from normal to cancerous growth and keep it there.

Now more ambitious questions arose. Might the cell itself have such genes? Might all cancers arise from the wayward actions of genes? Can the complexities of human cancer be reduced to the chemical vocabulary of DNA? Harold Varmus and I began to pursue these questions not long after he joined me in San Francisco in 1970. We were motivated in large part by curiosity about where the *SRC* gene of Rous sarcoma virus might have originated. Two considerations gave rise to our curiosity.

First, there was an evolutionary puzzle. The *SRC* gene apparently makes no contribution to the welfare of Rous sarcoma virus: it is fully dispensable, without harm to the virus.[26] Why then is it there? Might *SRC* have originated as a cellular gene and later found its way into the Rous sarcoma virus by means of a molecular accident? Second, there was the "Oncogene Hypothesis," put forth by Robert Huebner and George Todaro in 1969, which attributed all cancer to the activation of oncogenes intrinsic to cells. Huebner and Todaro imagined that these genes had been implanted into vertebrate germ lines by viral infection eons ago and now lay silent unless aroused by a carcinogen. Perhaps *SRC* embodied one of these hypothetical "enemies within." The Oncogene Hypothesis was largely fantasy, but it had heuristic value nonetheless.

Both the evolutionary puzzle and the Oncogene Hypothesis suggested that it might be profitable to search for *SRC* in the DNA of normal cells. I for one failed to foresee the eventual outcome. The work began when Harold suggested how we could prepare a molecular probe that would sense the presence of *SRC* amidst a welter of other genes. From that point, it required the better part of four years before we reached the conclusion that vertebrate cells do indeed carry a version of *SRC*. The first revealing experiments with chicken DNA were performed by our colleague Dominique Stehelin, a talented young French scientist who had joined Harold and me for further training. Here is how he later described his reaction to the first successful results:

"The intensity of the emotion I experienced and the intellectual clarity induced by the situation at that moment were very special . . . The fantastic results came out in the night of Saturday, October 26th, 1974: Normal DNA contained sequences related to the *SRC* gene of the transforming virus . . . I suspect that few have the privilege of enjoying such a moment when one is intensely and profoundly aware that a major step forward in Science has been made, and that one has contributed to it."[27]

Where was I at the moment of magic? Dominique and I often

shared an evening meal at a local sandwich shop, but by the time the data chattered out of a radiation detector that October evening, Dominique was alone in the laboratory, whereas I was at home and possibly in bed (it was Saturday, after all). Which calls to mind a vignette involving the particle physicist and Nobel laureate Carlo Rubbia. After one late-evening conference with a graduate student, Rubbia is reputed to have said: "Now I go home to sleep, and you go back to work." Students who read this anecdote with pain can now think of Dominique Stehelin and know the potential nature of his satisfaction: a moment of magic not shared with anyone else, particularly a mentor.

But was it a cellular or viral gene that Dominique had detected in chicken DNA? We first sought evidence that the gene had been conserved from one species to another through the course of evolution, which would be typical of a cellular gene but most unlikely for a viral gene. Harold recruited Allan Wilson, an evolutionary biologist at the University of California, Berkeley, as our advisor in this effort. (Wilson died some years later of leukemia.) Allan turned our attention to the ratites (ostrich, emu, rhea, and cassowary), considered to be among the most primitive of surviving birds and thus the ones most diverged from chickens. Accordingly, detection of *SRC* in ratites would be especially telling.

We had some difficulty in locating a zoo that would part with specimens for the purposes of research. But eventually, we obtained a newly hatched emu, already the size of a full-grown chicken. Soon thereafter, a full grown and recently deceased rhea arrived, frozen in dry ice. The rhea corpse yielded up DNA readily enough. But that baby emu was another matter. The creature proved so charming that no one among us would perform the necessary sacrifice. So a university veterinarian was pressed into service and Dominique was able to complete his survey, showing that the cellular version of *SRC* was present wherever he looked among birds.

Our first published report of *SRC* in normal cells appeared a little more than a year after that exciting night in October 1974. We were not blind to the potential significance of our findings, as shown by the final sentence of that report: "We are testing the possibility that [the *SRC* in normal cells is] involved in the normal regulation of cell growth and development or in the transformation of cell behavior [to a cancerous state] by physical, chemical or viral agents."[28] Indeed, the manuscript that we submitted for publication contained a claim that *SRC* was activated in cells transformed to cancerous growth by a chemical, in accord with the original Oncogene Hypothesis. But a referee at the journal *Nature* argued that we had pressed our analytical technique to its limits and recommended caution. We heeded the advice, removed the claim from the version of the manuscript that was to be published, and were soon glad of that when the original observation proved to be erroneous: in reality, *SRC* was active in both the normal and cancerous cells. We had indeed pressed the analytical technique too far.

What of mammals? In a first try, Dominique had failed to detect *SRC* in mammalian DNAs. We were moved to revisit the issue, however, when a test intended only to measure background "noise" in our assay turned up *SRC* in normal mouse cells. Another of our young colleagues, Deborah Spector, pursued this clue and soon had evidence that humans, mice, cows, and fish could be added to our catalogue of creatures with a *SRC* gene. Or so we believed. But there was vigorous skepticism in other quarters. When our detection of *SRC* in human DNA was first reported at a major symposium, it was greeted with the incredulous response: "Are you trying to tell us that a chicken gene is also in humans?" I was flabbergasted by this biological naivete on the part of accomplished scientists. Had Darwin labored in vain?

The skepticism was fully dispelled only with the advent of recombinant DNA. Then it became possible to show decisively that *SRC* is indeed a normal cellular gene, grafted into the virus of Peyton Rous by an accident of nature during the course of viral propagation. Now we faced a new question. Was *SRC* a curiosity or an archetype? Had the oncogenes of other retroviruses also originated from cellular genes? Diana Sheiness pursued this question with a chicken retrovirus known as MC29, which had attracted our attention because it causes carcinomas—the most common form of malignancy. Through exceptionally laborious work, Diana produced a molecular identification of an oncogene in the genome of MC29 that was soon dubbed *MYC*, and found that this gene also had a counterpart in normal cells. She finished in a dead heat with Dominique, who by then had his own laboratory in France and had succeeded in identifying cellular counterparts for three retroviral oncogenes, *MYC* among them. The example of *SRC* was not an exotic anomaly. It was an archetype.

We were excited. But others were not. We found our efforts to publish the story of *MYC* rebuffed by two referees for a fashionable journal. They created the literary equivalent of Scylla and Charybdis. One argued that the story of *MYC* was mundane, that the genesis of all retroviral oncogenes from normal cells had become self-evident with the results for *SRC* alone. The other referee argued that we could not claim to have generalized the principle until we could provide yet another example: *MYC* itself would not suffice.

Imagine how bored the first referee must have been, and how pleased the second, as additional examples tumbled out. The DNA of vertebrates contains many genes that can be pirated into retroviruses, there to become oncogenes. We call these cellular genes "proto-oncogenes" because each has the potential to become an oncogene in a virus. Meanwhile, additional means to uncover proto-oncogenes have been found and these have expanded the repertoire to one hundred or more. Each proto-oncogene can be found in many different species, from humankind to sea urchins, arrayed across 1 billion years of evolution. This conservation suggested to us that each of these genes serves a vital purpose for the organisms in which it is found.

From the outset, it seemed unlikely that evolution installed proto-oncogenes in our cells to cause cancer. These genes must have more benevolent functions (as we now know to be the case). Why then does their transfer into retroviruses give rise to oncogenes? The answer lies in the elaborate molecular gymnastics by which proto-oncogenes are pirated into the genomes of retroviruses. During the pirating, proto-oncogenes suffer damage that can convert them to oncogenes, from Dr. Jekyll to Mr. Hyde. So the virus is an inadvertent pirate; the booty is a cellular gene with the potential to become a cancer gene; and the conversion to oncogene is one of those accidents of nature that reveal deep truths to scientists.[29]

The discovery of proto-oncogenes inspired a further hope. Perhaps these genes exemplify a genetic keyboard on which all manner of car-

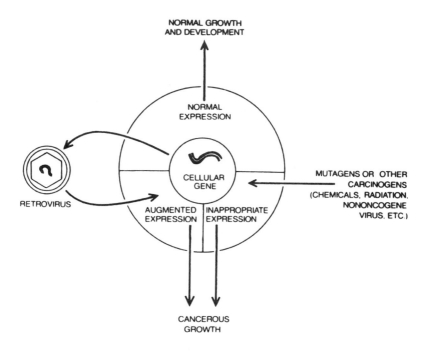

A genetic paradigm for cancer. As conceived by the author and drawn by Bunji Tagawa, 1982. An early portrayal of how cellular proto-oncogenes might be coopted into becoming oncogenes in either viruses or cells. (Reproduced by permission of the estate of the artist.)

cinogens might play. Any agent that can damage a proto-oncogene might give rise to an oncogene, even if the damage occurred without the gene ever leaving the cell, without the gene ever confronting a virus. In this view, proto-oncogenes are precursors to cancer genes within our cells, and damage to genes becomes the underpinning of all cancers, including those not caused by viruses.

The scheme would not have sat well with Peyton Rous. He made his sentiments about mutations and cancer very clear in his Nobel Lecture of 1966: "No inkling has been found . . . of what happens in a cell when it becomes neoplastic, and how this state of affairs is passed on when it multiplies . . . A favorite explanation has been that [carcinogens] cause alterations in the genes of cells of the body, somatic mutations as these are termed. But numerous facts, when taken together, decisively exclude this supposition."[30] In this case, Rous was dead wrong—so much for the infallibility of Nobel laureates.

Right or wrong, Rous could be a bulldog in debate: "A hypothesis is best known by its fruits. What have been those [fruits] of the somatic mutation hypothesis? It has resulted in no good thing as concerns the cancer problem, but in much that is bad . . . Most serious of all the results of the somatic mutation hypothesis has been its effect on research workers. It acts as a tranquilizer on those who believe in it."[31]

Rous had two reasons for his opposition to the "somatic mutation hypothesis" of cancer. First, he argued that many carcinogens are not demonstrably mutagens, citing tumor viruses as a prime example. Here, Rous was wrong again, although that would not become apparent for several decades after his diatribe. Second, Rous found the genetic view of cancer "fatalistic" and feared that it might discourage efforts to cure cancer through chemotherapy—a notable miscalculation of human hope and persistence. Rous's dialectic notwithstanding, it was not long before damage to proto-oncogenes had been implicated in the genesis of human tumors. I will give two examples.

The first example returns us to cytogenetics. The discovery of proto-oncogenes led in turn to a solution for the molecular conundrum of chromosomal translocations, as first encountered in the Philadelphia Chromosome. The breakthrough came from a study of lymphatic tumors in mice and humans. In both instances, the tumors contain characteristic translocations, and these disturb the proto-oncogene *MYC*, moving it from one chromosome to another. As a result, the gene is switched on inappropriately with disastrous consequences.

Soon thereafter, it was found that the translocation represented by the Philadelphia Chromosome mangles another proto-oncogene, known as *ABL* and also first encountered in a retrovirus; the mangling creates what is in effect a cancer gene. These discoveries with *MYC* and *ABL* were greatly aided by the fact that both genes were already known to us through the study of retroviral oncogenes. By now, it has become axiomatic that most of the many chromosomal translocations associated with human tumors convert proto-oncogenes to oncogenes.

A very different line of inquiry provided dramatic additional evidence to incriminate proto-oncogenes in tumorigenesis. In 1979, Chaiho Shih and Robert Weinberg at the Massachusetts Institute for Technology took DNA from cells that had been converted to cancerous growth by a chemical carcinogen and introduced this DNA into normal cells. A few of the recipient cells converted to cancerous growth: the DNA had transferred an active cancer gene. Here was a strong boost to the view that DNA is indeed the target for chemical carcinogens, and that the activity of genes is what drives the malignant behavior of cancer cells.

When the same procedure was repeated with DNA from human tumors, a similar result was sometimes obtained. In those instances, the tumor DNA apparently contained a biologically active cancer gene. Shih and Weinberg at first thought that they might have uncovered a new form of cancer gene. But when the gene in question was isolated from tumor cells, it proved to be a mutant version of a proto-oncogene with the name *RAS*, once again a culprit already known to us from the study of retroviruses—even the mutation in the tumor gene was identical to one found in the *RAS* oncogenes of retroviruses. It now seemed likely that *SRC* was indeed an archetype for a whole battery of genes commonly involved in the genesis of cancer.

We now know that many human tumors contain mutations in *RAS* proto-oncogenes (the human genome contains three closely related versions of *RAS*). Some of the more prevalent examples include cancers of the colon, lung, pancreas, and bladder. The damage in these genes is far more subtle than that represented by chromosomal translocations. A single letter in the genetic code has been altered, a change that geneticists call a "point mutation." Despite the simplicity of the change, the result is a highly potent cancer gene.

The initial catalogue of proto-oncogenes was developed through the study of retroviruses. But the molecular dissection of human tumors has since added many dozens to that catalogue—the tally now exceeds one hundred. As the tally grew, so did the tie between proto-oncogenes and cancer. It now appears that most if not all human cancers contain damage to one or another proto-oncogene. The damage takes three general forms: point mutations of the sort found in *RAS*;

scrambling of genes among different chromosomes, by translocation; and an overgrowth of genes that we call amplification. In each instance, the damage somehow unleashes the function of the gene, so that it runs relentlessly and perturbs the behavior of cells.

Consider in retrospect what retroviruses have done for us. Piracy of proto-oncogenes by retroviruses is an accident of nature, serving no purpose for the virus. But the event has profound implications for cancer research. In an extraordinary act of unintentional benevolence, retroviruses have brought to view cellular genes whose activities may be vital to many forms of carcinogenesis. It might have required many decades more to find these genes by other means among the morass of the human genome. Instead, we have the genes made manifest in retroviruses, excerpted from the genome and made available for our closest scrutiny.

Tumor Suppressor Genes

The discovery of proto-oncogenes and oncogenes gave substance to the excess of chromosomal function predicted by Theodor Boveri. But what of the chromosomal deficiencies he also had imagined? The first hint of these came from the work of Henry Harris and his colleagues in the 1960s. Harris had helped perfect a procedure for fusing two cells together. This causes the genetic dowries of the two parent cells to commingle in the resulting hybrid cell, which then continues to proliferate as if nothing were amiss.

Harris and his collaborators then found that the fusion of normal and cancer cells often suppresses the malignant behavior of cancer cells. The hybrid cells grow normally rather than as cancer cells. From this work, it was inferred that cancer cells might be defective in genes that are required for the regulation of cellular proliferation and other behavior. Fusion with a normal cell restores the necessary genes and thus suppresses cancerous growth.

The hypothetical genes became known as "tumor suppressor genes," and their defectiveness in cancer cells evokes the chromosomal deficiencies first imagined by Boveri. But the experiments of Harris did not lead us to individual genes. Instead, the route to the isolation of tumor suppressor genes lay through the study of inherited cancer, the last of our convergent themes.

Inherited Cancer

In 1866, the French neurosurgeon and anthropologist Paul Broca sketched the pedigree of his wife's family and discerned a hereditary predisposition to breast cancer.[32] In the decades that followed Broca, descriptions of inherited cancers continued to appear sporadically. But at least two difficulties kept these descriptions from having any immediate effect on how scientists thought about cancer.

First, examples of familial cancer are rare. The nonfamilial tumors that dominate clinical experience might be entirely different in their origins, although I for one have always found that reasoning to be specious. The inheritance of cancer, no matter how rare, clearly demonstrates that genes are capable of contributing to tumorigenesis. Why should the same and similar genes not also be involved in nonfamilial cancer?

Second, the patterns in which cancer is inherited are often confusing. In some instances, whole generations are skipped by the disease. In others, more than one type of tumor is inherited, but the pattern of the inheritance is not predictable. Biologists do not deal well with such disorder, so the inheritance of cancer was slow to enter the mainstream of thought.

There was a special case, however, in which there could be no doubt about what was happening. Cells possess a variety of devices that can repair DNA after it has been erroneously copied or otherwise damaged. On occasion, one or another of these devices is congenitally defective. The first example to become clear was a disease known as xeroderma pigmentosum, an inherited malady that features extreme sensitivity to sunlight and a frightening predisposition to skin cancer. Unless rigorously shielded from sunlight, afflicted individuals develop life-threatening cancer by early adulthood.

While I was still in medical school, James Cleaver recognized xeroderma pigmentosum as a deficiency in the repair of DNA damage caused by ultraviolet radiation (the carcinogenic component of sunlight). Defects in any of several genes can give rise to xeroderma pigmentosum. These defects are not directly tumorigenic. But they are nevertheless disastrous: failure to repair DNA allows mutations to accumulate far beyond their usual frequency, and in due course, mutations will occur that are directly tumorigenic.[33] With this scheme in view, the mutational theory of cancer took on new luster. Here was a clear indication that damaged DNA carried the risk of cancer. I have been a believer in the somatic mutation hypothesis of cancer ever since.

Familial cancers not based on defective DNA repair were less transparent. The first of these to come under close scrutiny was retinoblastoma, a relatively rare tumor of the retina restricted to children under the age of five. Perhaps 30 percent of retinoblastomas occur in an inherited pattern. An early clue to how these tumors might be inherited came from cytogenetics.

Some retinoblastomas contain a chromosome (number 13) that has lost part of its DNA, a lesion that is called "deletion" and that is sometimes large enough to be visible through a microscope. In families that are transmitting retinoblastoma from one generation to the next, the chromosomal deletion is always inherited in concert with the predisposition to cancer, as if it might in some way be responsible for starting the tumorigenic process.

The recognition of this deletion provided two vital pieces of information. First, it suggested that the inherited tumors suffered from a genetic deficiency, a loss of some vital function much as envisioned from the experiments with cell fusion. And second, it provided a location for that deficiency on a specific chromosome, a place for molecular biologists to begin digging. In 1986, the digging reached pay dirt: the gene affected by the deletions was isolated and dubbed the "retinoblastoma gene" (or *RB1* in formal nomenclature). From this work, it became clear that the same gene is affected in both inherited tumors and the slightly more common retinoblastomas that occur without inheritance. Deficiencies in *RB1* have since been implicated in a variety of other tumors as well.

Evidence has now been obtained for the participation of tumor suppressor genes in most if not all forms of human cancer. In each instance, the tumor cells have become deficient in the function of the tumor suppressor gene. A rapidly mounting number of these genes has been identified. Their abundance may well be similar to that of proto-oncogenes.

Most inherited cancers are based on deficiencies of tumor suppressor genes. In contrast, familial tumors attributable to mutant proto-oncogenes appear to be rare, perhaps because such mutations are often lethal to the embryo.[34] Bear in mind, however, that direct inheritance of cancer is unusual. Most cancer genes and most cancers arise from DNA damage incurred during our postnatal lives. The damaged genes are not inherited, but instead die with the individual in whom they arose.

The Genetic Paradigm for Cancer

We have arrived at the confluence of our five disparate themes. The malign behavior of the cancer cell is heritable because it is rooted in the genes of the cell. Genetic targets for the mutagenicity of carcinogens and the mangling action of chromosomal damage have been identified—proto-oncogenes and tumor suppressor genes. The cancer genes of viruses and the inherited elements of congenital cancer have engendered a comprehensive view of tumorigenesis. We have come to understand the genesis of cancer as a protracted and stepwise process, a sequence of mishaps that we believe are largely genetic. We have developed a genetic paradigm that unites all of cancer under one roof.

Genetic portraits of human tumors exemplify the paradigm. Virtually every human tumor that has been properly examined contains a combination of lesions in proto-oncogenes and tumor suppressor genes. These combinations appear to embody the multiple steps required to produce a malignant tumor. Each individual lesion adds insult to injury, the eventual sum being a malignant tumor. The catalogues of genetic lesions in cancer cells now available to us are astonishing. Less than twenty years ago, we knew nothing of the lesions and had no means by which to find them.

Proto-oncogenes cause trouble only when they do something they should not, whereas tumor suppressor genes are problematic only when defective or lost. These are diametrically opposite maladies, yet they play cooperatively on the cell to produce a single outcome—cancer. How does this happen?

The behavior of cells is governed by an elaborate network of molecular interactions that resembles electrical circuitry. Some portions of this circuitry mobilize the cell to necessary actions, such as proliferation, migration, differentiation, and other behavior required to create and maintain the structure and function of individual tissues. Proto-oncogenes represent switches in this part of the circuitry. The damage to proto-oncogenes in cancer cells creates molecular short-circuits: the network now signals relentlessly, driving the cell to unwanted actions. (Another useful simile is that of a jammed accelerator.) Other portions of the circuitry bridle the actions of cells. In this part of the network, tumor suppressor genes are switches, and inactivation of these genes deprives the cell of bridles, unleashing the cell to unwanted actions. (Here, the alternative simile would be that of a defective brake.)

The reduction of cancer to its genetic essentials is a source of pride and gratification for biomedical scientists. But their achievement was anticipated by an artist. In 1934, Diego Rivera painted an expansive mural in the Palace of Fine Arts of Mexico City entitled *Man, Controller of the Universe*. He had painted the same mural previously in the newly constructed Rockefeller Center of New York City (albeit with a different title—*Man at the Crossroads*), but that version had been destroyed after Rivera refused to remove an image of Lenin.[35] At the heart of the mural is a fanciful portrayal of the apparatus that facilitates chromosomal replication, and the apparatus is in turn gripped by a robust human hand. Rivera was unusual among artists in his strong belief that science and technology offer the greatest hope for the future welfare of humankind. The grip of that human hand exemplified his faith that we would some day understand the machinery of chromosomal replication and be able to turn that understanding to our advantage. That day now appears imminent.

Malignancy

Norman Mailer once captured the complexity of cancer: "None of these doctors has a feel for cancer . . . The way I see the matter, it's a circuit of illness with two switches . . . Two terrible things have to happen before the crud can get its start. The first cocks the trigger. The other fires it. I've been walking around with the trigger cocked for forty-five years."[36] The speaker here was a smoker who died of lung cancer four pages later in Mailer's novel *Tough Guys Don't Dance*. Mailer's conservative estimate of two "triggers" has since been revised upward for most cancers, but otherwise, the imagery is on target.

The multiple genetic events that contribute to tumorigenesis are thought to confer incremental properties that together create a malig-

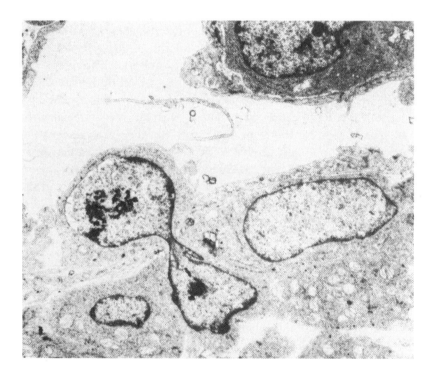

Metastasis. An electron micrograph has caught a cancer cell squeezing into the interior of a blood vessel. The squeeze has deformed the cell into an hourglass shape. One half is already within the lumen of the vessel, the other remains in the tissue surrounding the vessel. Left undisturbed, the cell would have eventually entered the bloodstream and spread to some remote part of the body. (Reproduced by permission of David Prescott.)

nant cell. For example, an emerging cancer cell might independently acquire capabilities for extended proliferation, for invasion into and migration through adjacent tissue, for penetration of lymph and blood vessels, and for spreading through the body. These are all properties that distinguish a malignant cancer from a benign tumor. We do not know the details of how this all happens. But we are reasonably certain that, some day soon, we should be able to assign distinct steps in tumorigenesis to individual genes.

It has been estimated that every gene in our DNA is damaged some 10 billion times in a lifetime. Yet the rate at which mutations arise is far lower, a tribute to the efficiency with which cells repair DNA. Given that efficiency, why do any of our cells ever accumulate the catastrophic combination of mutations required to generate a malignant cell? The answer to this long-vexing question is now in hand and represents an unexpected twist in the story. As cells reproduce, they monitor themselves for the completion of crucial events, such as the replication of DNA, repair of mutations, and construction of the apparatus required for cell division. If all is not well, a feedback device brings the reproductive process to a temporary halt, buying time for defects to be remedied. That failing, the cell can destroy itself by a form of suicide known as "apoptosis" in order to avoid becoming an outlaw.[37]

Some of the genetic damage in cancer cells cripples either the failsafe device itself or the capacity for self-destruction, allowing cells to be sloppier and, thus, to accumulate mutations that would otherwise not have survived. In other words, certain kinds of mutations can beget many more, facilitating the progression toward malignancy. The same genetic sloppiness also accounts in part for the relative ease with which cancer cells can become resistant to therapies.

The ability of cells to deliberately kill themselves came as a surprise when it was first discovered. But the capability is widespread in nature and plays a vital role in the sculpting of organs during development. For example, the vertebrate brain begins life with a large surfeit of cells, but many of these systematically destroy themselves as the brain matures and they become superfluous. The same cellular talent for self-destruction has been adapted to serve our intrinsic defenses against cancer.

Practical Implications

What might the genetic paradigm offer toward the control of cancer? It is too early to give a decisive answer to this question, but there are reasons to hope that the exploration of genes will eventually improve our ability to prevent, detect, classify, and treat cancer.

We can prevent cancer by reducing exposure to the external causes of the disease, and by intervening in the inherited risks of cancer. The genetic paradigm for cancer provides ways to strengthen both of these strategies. First, characterization of the damage in the genes of cancer cells may provide a new way to identify causes of cancer. This prospect can be dramatized with what we now know about skin cancer. Most such cancers contain a damaged version of a tumor suppressor gene known as *TSP53*. The chemical nature of the damage is characteristic of what happens to DNA when it is exposed to ultraviolet radiation. So even if we did not know from previous epidemiological studies that sunlight is the principal cause of skin cancer, we could strongly suspect the cause from the damage in *TSP53*. For cancers whose causes have yet to be established, we hope to reason "backward" from the nature of the genetic damage in these cancers to the nature of what caused the damage and, thus, the cancer.

Second, genetic screening can be used to identify individuals who have inherited an increased susceptibility to cancer. But having such knowledge can be a mixed blessing. In some instances, such as hereditary melanoma of the skin, there is presently no means for intervention other than careful monitoring. Some established interventions are plagued with uncertainties—the use of anti-estrogens such as tomoxifen to deter breast cancer is a familiar example. And some interventions are draconian, yet only partially effective—prophylactic mastectomy to avoid inherited breast cancer leads this list.

The advent of genetic screening also confronts the practice of oncology with new dilemmas. Will genetic screening for cancer improve detection sufficiently to justify its use and expense? How will it fit into the changing landscape of the medical marketplace? What implications might it have for insurance and employment? Alert to these concerns, an advisory council to the National Institutes of Health has warned against the general use of genetic screening for susceptibility to cancer until more is known about the efficacy of the screening and its societal effects. And states have begun to legislate prohibitions against the use of genetic profiles by employers and insurers.[38]

Effective therapy of cancer is best assured by early detection of the disease. But we presently have screening techniques for only a modest number of human cancers, and several of these techniques remain beset with uncertainty. For example, there is continuing debate about whether the use of mammography has a beneficial effect on mortality from breast cancer, and disquiet over the unnecessary interventions occasioned by false positives in the test. Similarly, testing for prostate-specific antigen (PSA) may be detecting many tumors that are not life-threatening and would be better left alone, but we presently have no way to recognize that class of tumors.

Genetic screening may provide a helping hand with these uncertainties. For example, human excretions such as sputum, breast fluid, urine, and feces carry cells shed from the interior of the body. It is now possible to screen those cells for genetic damage that signifies the presence of cancer. This "genetic cytology" may be both more sensitive and more revealing than currently established techniques such as X-rays, scans, endoscopy, and microscopy. A retrospective look at the death of the American statesman Hubert Humphrey can illustrate these advantages.

Hubert Humphrey died of bladder cancer. Scientists have recently used genetic cytology to examine urine and tumor tissue taken from Humphrey and preserved after his death. They found that the bladder cancer could have been detected six years earlier than it was had genetic screening of cells in the urine been available; and that the analysis would have prompted immediate, aggressive therapy—in all likelihood, curing Humphrey of his cancer. Given the rigors of therapy, he might well have decided not to run for the presidency against Richard Nixon in 1968—the year that his cancer could first have been detected by molecular cytology.[39]

Does the genetic paradigm promise new therapies for cancer? It is unlikely that we will be able to repair or replace the damaged genes of cancer cells in the foreseeable future: we have not yet learned how to operate on the DNA of living human cells with the necessary accuracy and efficiency. There are other genetic strategies that aim to switch off oncogenes in a direct and specific manner. But these too are far from realization.

If we focus on the protein handmaidens of genes, however, we can see more cause for hope. Given sufficient information about how these proteins act, we should be able to direct our therapies accordingly. In the case of proteins encoded by mutant proto-oncogenes, we seek ways to interdict the function of those proteins. The hope is to develop magic bullets of the sort first envisioned by Paul Ehrlich for bacteria, but directed instead at cancer cells. Targeting abnormal proteins in cancer may provide a way to avoid the toxicity for normal cells that engenders the noxious, sometimes life-threatening side effects of many current cancer therapies.

We can point to two promising examples, both involving abnormalities of proto-oncogenes. One is an agent known as Herceptin, which attacks a protein produced in abnormal abundance on the surface of approximately 30 percent of metastatic breast cancer cells. Herceptin has proven to be a valuable adjunct to the conventional therapy of breast cancer, but it is not curative. The other example is Gleevec, a drug aimed at the renegade chemical activity spawned by the Philadelphia Chromosome in chronic myelogenous leukemia. Gleevec has demonstrated remarkable efficacy in the first phase of the leukemia, when the disease is relatively indolent; but it has been disappointing in the treatment of the later, highly aggressive phase of the disease, in part because the cancer cells quickly develop genetic resistance to the action of the drug.

In the case of proteins inactivated by mutations in tumor suppressor genes, we seek ways to revive the proteins or provide alternatives to their activities. The prospects here are probably less immediate than those for intervention against mutant proto-oncogenes.

We are also beginning to learn how genetic profiles of cancer cells can be used in the management of cancer. These profiles can be obtained in two different ways: by looking directly at DNA for abnormalities associated with cancer; and by surveying the expression of many genes for changes in tumor cells. It is already apparent that these tactics will be useful in categorizing tumors and predicting their out-come. In addition, there is hope that genetic profiles will eventually be used to choose the most effective therapy for individual cancers. Just as we presently base the choice of antibiotic therapy on the specific sensitivities of the infectious agent, the treatment of every cancer may someday be individualized and tailor-made, according to the inventory of genetic lesions in the cancer. The largest impediment to that advance may prove to be its cost; much will depend on how many different genetic fingerprints there might be for any given form of cancer, and thus how diversified the tailor-made therapies might have to be.

No single therapy for cancer, no matter how specific and elegant, is likely to become a panacea. We must deal with a large variety of damaged genes whose actions present great functional diversity. We shall also have to cope with the genetic sloppiness of cancer cells that can bring additional cancer genes into play as treatment proceeds, and that can create resistance to therapeutic agents during treatment. In 1983, a prominent figure in American cancer research told the *New York Times* that "scientists should learn how to manipulate oncogenes to protect or treat patients within the next five years." The prediction has not been vindicated. The words ring hollow now, except as a cautionary tale.

Lessons

The genetic paradigm has provided a powerful view of cancer. The seemingly countless causes of cancer—tobacco, sunlight, asbestos, chemicals, viruses, and many others—may all work in a single way, by playing on a genetic keyboard, by damaging a few of the genes in our DNA. An enemy has been found, and we are beginning to understand its lines of attack.

The story of cancer research in our time embodies a great truth about scientific discovery. Peyton Rous isolated his virus from chickens, beasts not renowned for glamour. Yet the chicken virus isolated by Peyton Rous sired a remarkable lineage of discovery, replete with Nobel Prizes for five individuals.

The virus itself opened a new frontier in the search for causes of cancer, then served as the vehicle for additional discoveries of great consequence. These included reverse transcriptase, upender of genetic dogma; the viral oncogene *SRC*, the first explicit example of a cancer gene; proto-oncogenes, the first glimpse of a genetic keyboard for carcinogenesis; and the protein product of *SRC*, which provided the first example of a chemical reaction that can propel cancerous growth. All of this from a virus that, at the time of its discovery, was not deemed relevant to human cancer, all of this from the humble chicken.

Here is a familiar but oft-neglected lesson. The proper conduct of science lies in the pursuit of nature's puzzles, wherever they may lead. We cannot prejudge the utility of any scholarship; we can only ask that it be sound. We cannot always assault the great problems of biology at will. We must remain alert to nature's clues and seize on them whenever and wherever they may appear. H. G. Wells understood this lesson well: "The motive that will conquer cancer will not be pity nor horror; it will be curiosity to know how and why . . . Pity never made a good doctor, love never made a good poet. Desire for service never made a discovery."[40]

In 1978, Susan Sontag described cancer as "overlaid with mystification, . . . a triumphant mutation, . . . charged with the fantasy of inescapable fatality, . . . a scandalous subject for poetry."[41] Now the force of science has taken some of the sting from those words. The mystification is in retreat, the triumphant mutation has been exposed, we see new ways by which to confront that inescapable fatality, and there is even reason for poetry.

> But the comfort is
> In the covenant
> We may get control
> If not of the whole
> Of at least some part
> Where not too immense,
> So by craft or art
> We can give the part
> Wholeness in a sense.[42]

CHAPTER 5

Paradoxical Strife

[It is possible] to believe that the age of science and technology is the beginning of the end for humanity; that the idea of great progress is a delusion, along with the idea that the truth will ultimately be known; that there is nothing good or desirable about scientific knowledge and that mankind, in seeking it, is falling into a trap.

—*Ludwig Wittgenstein*, Culture and Value

Faust lisant by Salvador Dalí, date uncertain. (Reproduced by permission of CFM Gallery, New York City.)

The fruits of science have vastly improved human understanding and welfare. We have found and decoded the molecular apparatus of our inheritance and will eventually be able to reconfigure that apparatus if we so choose. We have traced the origins of life to an unimaginably early time, more than 3 billion years ago. We have enumerated the immense diversity of life—more than 1 million species accounted for, but millions more left to find, should they survive the onslaughts of Homo sapiens. We have laid bare a genetic kinship between ourselves and all of the other creatures that inhabit the earth, a kinship that inspires respect for all things living, yet that some deny because it offends their religious beliefs.[1] We have broken matter into smaller and smaller pieces, only to learn that the ultimate components are not particles, as Democritus and we once imagined, but subatomic fields of energy. We have reached so far into space that only future generations will be able to decipher all that we recorded. We have cured and prevented diseases that once killed millions every year.

These are great successes that ennoble us all. But they and their kind have inspired an unexpected disaffection among the general public and within our government. We have been compelled to reconsider the ways in which science proceeds, the benefits and stresses that it brings, and the means by which it can be sustained. In this final chapter, I recount some of the ways that the disaffection with science has touched my life and career. I wish to show in a substantive way why I felt some compunction to write this book. I will speak from the vantage point of a biomedical scientist. But I will be illustrating problems that pertain to all of science.[2]

DNA in the Neighborhood

I first witnessed a substantive clash between science and the public in 1986, when the University of California, San Francisco (UCSF) sought to convert a large office building into medical research laboratories. The building is located in a residential neighborhood that is relatively affluent and educated. The university expected its new neighbors to be sympathetic to its mission and proud of its presence. Instead, they mounted a costly and enervating resistance that lasted more than five years, grew increasingly vitriolic as time passed, and concluded with a Pyrrhic victory for academe.

Our prospective neighbors—at least the activist minority among them—had nothing good to say about us.[3] They argued that we exude toxic wastes, infectious pathogens, and radioactivity; that we endanger the life and limb of all who come within reach—our own lives and limbs included, I suppose, a nuance that was lost on the opposition (as was the fact that we feel free to take our children to work with us); and that we create quagmires of traffic and are gluttons for parking space (arguably the risks that the neighborhood dreaded most of all).

In retrospect, the dispute was much more about incursions on private lives than about any fundamental disaffection with science. But the citizens who opposed the university demonized science as a tactic, exploiting misconceptions that alarmed and angered at least some of their audience in the neighborhood. The university in turn relied too heavily on the presumption that the benefits of science are both self-evident and innocuous. The episode served as a prelude for my personal consideration of the more substantive ways in which science and society have come into conflict.

Two vignettes from the fray dramatize the misconceptions that marked the discourse: an agitated citizen, suggesting in public forum that the manipulation of recombinant DNA at UCSF had accidentally engendered the AIDS virus (in reality, the virus almost certainly originated from chimpanzees in Africa); and an elderly denizen of the neighborhood, declaring over television her outrage that "those people are bringing DNA into my neighborhood," apparently unaware that there was already quite a bit of it there—900 trillion yards of it, to be exact, in each living resident.[4]

In all fairness, it should be added that fear of DNA was not at the time restricted to San Francisco. Mayor Alfred E. Velluci of Cambridge, Massachusetts, asked the president of the National Academy of

Sciences to investigate whether recently sighted aliens—"a strange, or-
ange-eyed creature" and a "hairy nine foot creature"—were "in any
way connected to recombinant DNA experiments taking place in the
New England area."[5] Mayor Velluci's animus toward academia, particularly Harvard University, was legendary—his campaign platforms
regularly called for conversion of Harvard's venerable inner "Yard"
into a public parking lot.

These views and others of similar tone fueled a reaction that
stopped the university dead in its tracks and fostered disillusionment
that has lingered well beyond the end of the struggle. A city official
eventually called the episode "one of the most tragic" in the history
of San Francisco, momentarily neglecting the city's record of catastrophic earthquakes.

The issue was eventually resolved by the Supreme Court of California. In finding for the university, the court sought to balance risk
against benefit, and invoked the need for public altruism. The court
chided the university for not having done its homework well
enough—they found the initial environmental impact report to be
deficient. But once that deficiency was rectified, the court said that it
did not wish to "shackle the scientific imagination" with unrealistic
standards; it acknowledged the inherent unpredictability of research
and its hazards; it even argued that the salutary nature of the university's mission mitigates the unpredictability, indeed, mitigates the
hazards themselves (which the court recognized as miniscule, in any
event).

The court had in its own way justified the daring of science. On
occasion, that daring can bring humankind face to face with the unknown, a confrontation that the citizenry of developed nations increasingly seem inclined to avoid. But rejection of the unknown carries hazards of its own:

> [Humankind] has always faced risks, whether in exploring un-
> charted territories or trying unfamiliar foods. If our recent success
> in conquering many malign forces of nature now leads us to seek the
> security of a world free from novel hazards, and if we forbid explo-
> ration of the new kind of unknown territory opened to us by sci-

ence, we shall not only be condemning ourselves to remain subject
to all the present, still unconquered risks; we shall be crushing one
of the most admirable expressions of the human spirit.[6]

For all the vaulting rhetoric, the university eventually capitulated to
reality and economics. The time consumed by legal action compelled a
strategic reconsideration of how the university might best use the
building, in part because conversion to laboratories had grown too
costly. So the building now houses administrators, epidemiologists,
sociologists, and children in day care. UCSF is building its new laboratories in an abandoned railroad yard, and it has worked very hard to
mitigate the concerns of the nearby neighborhoods. There has been no
appreciable opposition.

Neither UCSF nor its adversaries distinguished themselves in the
confrontation.[7] The university failed to explore in advance how the
neighborhood might respond to its plans for the building, proved in-
effectual when it finally took its case to the public, and managed inadvertently to appear deceptive. Its adversaries practiced what one com-
mentator called the "politics of intransigence."[8] Their objective was to
keep an institutional phalanx out of their neighborhood. In the process, however, they either misunderstood or misrepresented the risks
of biomedical science, they made no effort to genuinely understand
the university's purposes and practices, and they discounted the benefit that might accrue to the community at large from the university's
presence (most community organizations and businesses supported
UCSF in the dispute).

The extremity of the opposition was exemplified by one of its leaders, who commented that although space research is a good thing,
"you don't put the launching pad in the center of your residential
neighborhood"—not a particularly apt comparison to recombinant
DNA.[9] So it is that opposition to science usually derives from ignorance: the public fears the miniscule amounts of radioactivity used in
biomedical research, not knowing that their own bodies are naturally
radioactive; they fear the spread of recombinant DNA from one species to another, not knowing that genes have been migrating among
species of their own accord since the dawn of organized life; they fear

the presence of even miniscule amounts of synthetic chemicals in their environment, not knowing that the most pristine foodstuffs contain equally noxious materials, crafted by nature. It is not my intention to demean legitimate concerns about the hazards of technology. But the public assessment of those hazards is sometimes extreme.

Genetic Medicine

Science can interpenetrate and even destabilize society. The potential for disturbance can be illustrated by recent developments in human genetics. Put succinctly, "genetic medicine" is advancing rapidly upon us. We have come to realize that most if not all of our great maladies are grounded in our genetic dowry, and we have begun to act on that realization.

First in view were the rare hereditary diseases that arise from defects in single genes. Sickle-cell anemia, thalassemia, hemophilia, cystic fibrosis, Huntington's disease, and Tay-Sachs disease provide familiar examples. We have found the genetic underpinnings for many of the "single-gene" diseases and are closing fast on the others. The scope of the problem is substantial: the genetic defects responsible for such diseases are widely distributed in the human gene pool;[10] new reports of single-gene diseases reach the medical literature several times a month; more than one hundred inherited diseases afflict our retinas alone, and most of these are probably caused by defects in single genes.

Many of our more common ailments are also underlaid by genetic predispositions. Examples include atherosclerosis, hypertension, cancer, allergies, diabetes, Alzheimer's disease, schizophrenia, manic-depressive disease, and infections. We are not all equally susceptible to these maladies, whatever their immediate causes. The inherited roots of these diseases are complex, generally involving multiple genes, and not easily sorted out. But the roots are real, they have a deep influence on our susceptibility to disease, and they are rapidly becoming accessible to the experimentalist.

Take infectious disease as an example that might not come first to mind (unless you have read Chapter 3 carefully). Individuals vary greatly in their susceptibility to most infections and the diseases they cause, a variation whose bedrock is our genome. In general, only the most recent microbial invaders of our species enjoy uniform success in causing disease when they infect us—the deadly efficiency with which HIV causes AIDS is the latest reminder of this grim principle. Left to their own devices, the forces of evolution frequently reshape the genes of host and microbe to a less destructive interplay.[11] Thus, only one in every thousand individuals infected with the dreaded poliovirus develops neurological disease. We do not know the reasons for this, but genetic variation is surely among them. Other examples abound.

Our struggle against AIDS will probably be won by means of prevention, a time-honored strategy in our dealings with pestilence. But the struggle would be easier if we understood the rules that govern our response to the AIDS virus, and those rules are written in our genome.[12] Once we can make a profile of a person's genetic predisposition to disease, the practice of medicine will acquire previously unimaginable capabilities for prediction and prevention.

The dimensions of genetic medicine will continue to grow. Geneticists are now in pursuit of genes that influence personal traits such as temperament, sexual preference, even intelligence. This pursuit has enlivened the long-standing debate over determinism, the view that our genetic dowry is largely, perhaps even solely, responsible for what we become. Most biologists reject this view, believing instead that our fate is cast by an interplay between nature (our genes) and nurture (our environment). The influence of the genome on human behavior is probably best described as "probabilistic." The genome sets the stage and limits the possibilities, but much that would not be predictable follows in the course of human experience. "To be a human person means more than having a human genome, it means having a narrative identity of one's own."[13] Still, that "narrative identity" is not written onto a tabula rasa: "We are born knowing a thousand things we could not reinvent in a life time if we had to start from scratch."[14]

The virtually complete sequence of the 3 billion chemical units that compose the human genome was formally reported in February of 2001, an indelible landmark in the history of humankind. The techniques by which that sequence is scanned for the presence of individual genes remain imperfect, but the analysis to date suggests that no more than 1.5 percent of the sequence is actually devoted to genes, whose number is on the order of thirty thousand. Much simpler creatures, such as fruit flies and worms, can have a third or more of that number. How could a creature of such complexity and capability as Homo sapiens be wholly determined by only thirty thousand elements?

This seeming paradox inspired some extraordinary headlines ("Importance of DNA Diminished" is my favorite), but in reality, biologists know that the number of genes in itself tells us little about the complexities of life. The potential for combinatorial variation in the utilization of genes is vast: some genes specify more than one protein, the molecule that implements the instructions of the gene; some proteins have more than one function; proteins combine with one another in myriad ways to serve different purposes; and the expression of genes can be orchestrated into countless combinations, to achieve different ends. All in all, the number of potential outputs from such combinatorial schemes is immense: estimates range into the billions, complexity sufficient to account even for the glories of the human organism. So it remains credible that there is a genetic underpinning to all that we become, and that the underpinning will eventually be unveiled. But still, nurture remains in play.

And even nurture is not the end of influences. There are inherent biological variations that occur during growth and development, variations that are not dictated by genes. A splendid example has been provided by the first "cloning" of a cat. The donor of genetic material for the cloning was a calico female. The progeny had a calico coat too, as anticipated—all of the genes in the kitten had been derived from mom (there was no dad, of course—see below for more on cloning). But the pattern of coloring differed from that of the mother. The reason is that the cells responsible for skin pigment undertake a lengthy expedition early in life, migrating from one place to another in the developing embryo. The path of migration and its destination vary independently of genetic determinants. So the calico clone acquired a distinctive coat, and genetic determinism was dealt at least a modest blow.

Genetic Testing

It is now only a matter of time before we will be able to test for all substantive genetic predispositions to disease. The task will be large. We would probably examine millions of individuals each year. But place this in perspective. More than 5 billion medical laboratory tests are already performed in the United States annually. Genetic screening might increase that burden by no more than a few percent. This would not be an impossible or even impractical task should we choose to do it: we already perform more than 2.5 million genetic tests annually for Down syndrome alone.[15]

Genetic testing can lead to two forms of preventive intervention: elective abortion, when prenatal testing detects a heritable malady in the fetus; and recourse to behavioral and medical measures, implemented after birth. Whatever its objective, elective abortion remains repugnant to many and anathema to some. But when utilized, it can have a dramatic effect. Following the introduction of prenatal screening for the hereditary and devastating blood disease thalassemia, the frequency of afflicted newborns soon declined substantially in several

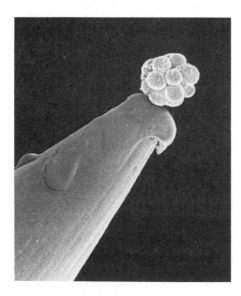

Human embryo. A human embryo three cell-divisions after fertilization, anchored on the tip of a glass micropipette. Magnification ca. 1,000-fold. (Photomicrograph by Dr. Yorgos Nikas. Courtesy of Science Photo Library.)

Mediterranean countries, where the natural incidence of the disease is relatively high and prevention a public priority.

In principle, diseases such as thalassemia could be eradicated from individual family trees by systematically coupling prenatal testing with elective abortion of afflicted fetuses, or by utilizing genetic testing after in vitro fertilization and discarding the embryos that are destined for disease (a procedure known as "preimplantation diagnosis"). Neither of these strategies is acceptable to those who oppose destruction of human embryos under any circumstances. We will encounter such opposition again when we take up the subject of human stem cell research.

Preventive medicine offers great promise. Current estimates suggest that more than 60 percent of all deaths in the United States are premature—that is, preventable. The advent of genetic testing will offer the chance to change that figure for the better, to alert those who should take special precautions against cardiovascular disease, diabetes, various cancers, or numerous other diseases whose frequency must be determined in part by inheritance, but can also be reduced by personal behavior (such as cessation of smoking) or medical intervention (such as the drugs that reduce serum cholesterol).

But the opportunity for prevention through genetic testing creates difficulties of its own. For example, more than twenty years ago, Sweden implemented screening of children for a genetic defect that greatly increases the risk that smokers will acquire the chronic lung disease emphysema. There seemed to be no earthly reason not to provide such seemingly harmless and helpful information. But being identified as having a "genetic defect" carried such a great stigma that the screening program was discontinued.[16]

Postnatal testing can also raise profound ethical issues. Huntington's disease, which killed folksinger Woody Guthrie, dramatizes these issues. For some time, it has been possible to screen for the defective gene that gives rise to Huntington's disease. If one of your parents has Huntington's disease, would you want to know at an early age that you are destined for this still incurable disease, or would you want to await the gradual onset of symptoms in middle age to announce your fate? And what of your children? If you are tested, their eventual fate may be immediately clarified. Should you tell them the news? Will they want to know? Moreover, they acquire the independent right to be tested at the age of eighteen and their testing will immediately reveal your own fate, even if you had earlier chosen not to be tested. Should they tell you the news? Will you want to know? And what of the physicians who first obtained the test results? How widely (if at all) are they obliged to share what they know? The courts are presently divided on this issue, some arguing for strict confidentiality, others arguing for a right, even an obligation, to share the information with individuals who might be at serious risk.[17] There can probably be no decisive answer to any of these questions until we have a way to prevent or cure Huntington's disease.

Medical research has produced more than nine hundred genetic tests for disease (although only a few are in regular use). And consumers appear generally disposed toward using these tests. In one recent study, 78 percent of women at high risk of breast cancer because of family history wanted genetic testing to ascertain their own risk. Even in the absence of any preventative intervention, individuals may want to know their risks. For example, 23 percent of individuals who had at least one parent with Alzheimer's disease said they would want to be tested for genetic predisposition to the disease, even though neither prevention nor treatment is presently available.

The prospects of genetic testing also pose economic issues to resolve. Will genetic screening be affordable under any circumstance, under any system of health care? Will it be cost effective—that is, will it improve the detection of susceptibility to disease sufficiently to justify its use and expense? These issues loom large because health-care providers are hard-pressed to deny a genetic test if it has demonstrable benefit. Currently, 97 percent of insurance claims for a genetic breast cancer test are being reimbursed in the United States (at the rate of $2,800 each), as are 60 percent of claims for a colon cancer test ($2,000 each). And in the end, will we be able to act on the information that genetic testing provides, or will we be frustrated again by our inadequacies in the prevention of disease, inadequacies that often stem from failures of personal and political will?[18]

The Social Consequences of Genetic Medicine

The need for confidentiality in genetic testing has inspired great anxiety. If potential employers and insurers were given access to the genetic profile of an individual, they would be unlikely to ignore warning signs written there. Indeed, there is concern that genetic testing might become a requirement for insurance, employment, even financial credit. Both the U.S. government and individual states have begun to enact legislation that prohibits access to the results of genetic testing without consent of the individual and that limits the use of such results in various ways. (As of this writing, thirty-nine states have laws to limit the use of genetic information by insurers, and approximately half have laws restricting the use of such information by employers.) But consent can be coerced by social and economic means, and existing legislation is imperfect. The problem remains unsolved.

There is no question that genetic testing has begun to penetrate the workplace. In February of 2001, the federal Equal Employment Opportunity Commission filed its first court action to challenge genetic testing by an employer. The action claimed that the Burlington Northern Santa Fe Railway Company had been performing genetic tests on employees who developed carpal tunnel syndrome, an ailment of the hand thought to be brought on by repetitive motion. The railway company was alleged to be seeking evidence of a genetic predisposition to the ailment (none is presently known), in the hope of avoiding claims for worker's compensation. Some employees reported being threatened with dismissal if they did not agree to the testing; others asserted that they had been asked for blood samples without explanation of how these would be used.[19]

Whatever the difficulties that it poses, genetic testing can be a blessing when applied suitably. Some years ago, the *San Francisco Examiner* carried a feature article about genetic testing, focused on cystic fibrosis. The article told of a couple who learned first that both were genetic carriers of the disease, then that the twins they had recently conceived would develop the disease after birth. After wrenching deliberation, the couple decided to continue the pregnancy.

Here is what the mother had to say during the last trimester of her pregnancy. "We will be sad parents, but parents who have had the chance to grieve for their loss. Parents who are informed and able to provide [their children] with a healthy and loving environment. Am I glad that I had genetic testing? Yes. I have a sense of what I am facing and I am ready to do my best for my children. Knowledge is indeed power."[20]

There is concern that genetic testing will refuel eugenics—the effort to retailor human heredity by any means possible.[21] Surely our society can find the wisdom and means by which to deal with that prospect. It is not eugenics that I fear. Instead, I share with the late Nobel laureate Max Perutz a different anxiety, the prospect of "a democracy so scared of science that it might accede to the shrill demands and intimidation by those who want termination of pregnancies to be banned, together with genetics and all its works."[22] In the United States, executive, legislative, and judicial action have all been used to obstruct the application of biomedical science, usually in opposition to genetic advances. There is no better example of this than what has been happening to research on human stem cells.

Stem Cell Research

Stem cells are the progenitors for all the tissues of the human body. In its earliest form, the human embryo is little more than a sheltered cluster of stem cells, each one capable in principle of producing every sort of cell found in the adult organism—in the argot of biology, the stem cells are "totipotent" (or more modestly, "pluripotent," to acknowledge that we do not know the exact point in development when some of the cells begin to limit their choices). As the embryo grows, individual stem cells proliferate and begin stepwise maturations that eventually produce the myriad types of cells that compose the adult. The process of this specialization is known as "differentiation." It is by no means clear, as yet, how the ultimate fate of each stem cell is determined. But the determination must be exquisitely balanced in order to achieve the functional integration of adult tissues.

It is now possible to isolate stem cells from human embryos and propagate them in the laboratory. There is hope that we can learn how to instruct stem cells to choose a particular fate depending upon our need—blood cells to treat deficiencies of the bone marrow, heart cells to repair the damage inflicted by heart attacks, pancreas cells to provide insulin for diabetics, nerve cells to ameliorate degenerative diseases of the brain or spinal cord injury, and others. This prospect has inspired great hope among individuals with ailments that for now are incurable. But the prospect has been held hostage by a political firestorm ignited by the laboratory procedures that are used to obtain stem cells.

Embryonic stem cells can be harvested from two very different sources: fetuses from elective abortions; and test-tube embryos, created by "in vitro fertilization." The prospect of exploiting either elective abortions or embryos produced by in vitro fertilization offends many individuals in our society and thus generates opposition to research on human stem cells. Much of the opposition comes from those who oppose abortion itself, and who extend their opposition to the destruction of embryos created in a test tube. These opponents wield great political influence. So research on human stem cells in the United States has been greatly constrained by federal prohibitions.

In contrast to the restrictive climate in the United States, the British parliament has approved both the use of human embryonic stem cells in research and the deliberate creation of test-tube embryos for that purpose. The embryos must be utilized within the first fourteen days of their creation or be discarded. That chronological divide is based on a biological view of when an embryo becomes an "individual." The embryos used for deriving stem cells from in vitro fertilization consist of fewer than two hundred cells with only primitive distinctions among them. Although loosely known as embryos, they are more properly termed "blastocysts." Whatever their name, they are far from being individuals. First, their cells have yet to form distinctive tissues—indeed, none of the cells has yet to be individualized. Second, if the blastocyst is divided into two halves, each half has the ability to develop normally to term after implantation into the uterus. Biologists view embryos as "individuals" only after this ability to "twin" has been lost.

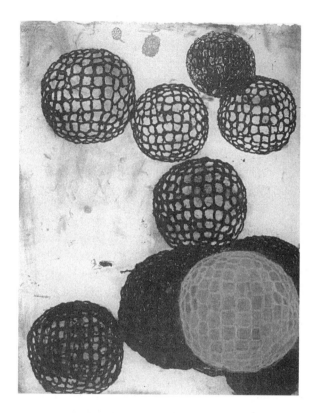

Morula III by Terry Winters, 1983–1984. "Morula" (from the
Latin *morus,* "mulberry") is the term used to describe the last
unstructured stage of vertebrate embryogenesis. The morula
matures into the blastocyst, in which stem cells first appear.
(Reproduced by permission of the artist and Universal Limited
Art Editions.)

The most readily available source of embryonic stem cells at present
are surplus embryos from fertility clinics.[23] The opposition to using
these embryos to derive stems cells seems paradoxical, since they are
likely to be discarded in the long term (on rare occasions, they are being "adopted" for use by infertile couples, but there is presently no
indication that this could become common enough to utilize all the
surplus embryos). If the opponents of stem cell research were intellec-

tually and ethically consistent, they would also oppose fertility clinics, but that is not a posture likely to find favor with the general public.

Some opponents have admitted as much. For example, the chief lobbyist for the National Conference of Catholic Bishops commented that "the church's moral opposition to *in vitro* fertilization [per se] has been pretty clear from the outset, but in terms of political action, we have to choose the issues that are raised for us."[24] In other words, fight only those battles that might be won. It is worth noting that when in vitro fertilization was first perfected more than two decades ago, it was widely opposed as unethical and inspired prohibitive legislation. But when its potential as a remedy for infertility became generally known, the opposition melted away.

The debate about stem cell research is compounded by the parallel furor over a procedure known as "reproductive cloning." In this procedure, the nucleus of an unfertilized human egg is replaced with that from an adult cell. The egg is then induced to proliferate into an early embryo, and this in turn can be implanted into the uterus of a surrogate mother, giving rise to an individual whose genetic dowry is derived entirely from the donor of the adult nucleus. The individual is a "clone."

Reproductive cloning has been performed successfully with frogs for several decades (without the need for a uterus, of course, and producing tadpoles but not mature frogs), with little attention from the general public. But in 1997, scientists in Edinburgh, Scotland, announced that they had produced a sheep named Dolly by means of reproductive cloning. Similar successes have been reported subsequently with other mammals. The upshot has been a hue and cry over the prospect of creating humans by reproductive cloning.

There are powerful scientific reasons to forbid the reproductive cloning of humans that cannot be denied even by those few who would otherwise advocate use of the technique. Among these reasons are the rarity with which reproductive cloning succeeds in mammals (it required 273 attempts to create Dolly, and similar inefficiency has been encountered in attempts with other mammals) and the high frequency of abnormalities among the individuals that result from the occasional "successes." Now deceased, Dolly herself was inexplicably overweight, suffered from arthritis at an early age, and aged prematurely. Other animals produced by reproductive cloning have congenital abnormalities of various organs, including the heart.

But the embryo produced in the first stages of reproductive cloning can be used for the alternative purpose of deriving human stem cells in what has been called "therapeutic cloning." This term has made its way into general parlance and I will use it here. But it is formally misleading, because the procedure does not result in a cloned individual; it results only in a harvest of stem cells.[25] Tissues derived from such stem cells would be immunologically compatible with the donor of the adult nucleus, thus circumventing rejection of the tissue when it is transplanted into the donor. So therapeutic cloning is imagined to be the ideal source of tissue for "regenerative medicine"—the provision of new pancreatic tissue to diabetics, for example, or of new nervous tissue to victims of various neurological diseases and spinal injury.

The apparent virtues of therapeutic cloning are presently outweighed by technical limitations, such as inefficiency and inordinate cost. More to the point, however, the procedure would require the sacrifice of a human blastocyst and thus is unacceptable to those who oppose the use of test-tube embryos for stem cell research.[26] Nevertheless, many scientists argue that research on therapeutic cloning should continue, in the hope of increasing the efficiency and reducing the cost, and of learning more about the early development of the human organism. In this view, legislation to prohibit reproductive cloning is justifiable, but it should be designed so that it does not interdict therapeutic cloning as well. This distinction could be achieved by prohibiting implantation into the uterus of any human embryo produced in therapeutic cloning.

Federal policy regarding stem cell research in the United States has been evolving erratically, so any précis of that policy is doomed to prompt obsolescence. But the policy has been consistent in the denial of federal funds to certain types of research. For a brief while under the administration of President Bill Clinton, federal support could be used for research on human stem cells only if the cells were originally derived with the support of private funds (the policy was enunciated but never implemented). More recently, President George W. Bush liberalized the policy in one regard, tightened it in another: federal funds could now be used for the study of any human stem cells that were derived before August 9, 2001; but all subsequent derivation of human stem cells and study of the newly derived cells would have to be supported entirely by nonfederal funds. The reader should not assume that this policy has stood (or will stand) the test of time.

The current demarcation between federal and private support for research on human stem cells seems subtle testimony to the power of the commercial ethos in the United States. Although the denial of federal funds is ostensibly based on ethical objections to the research, neither the executive nor the legislative branch of the U.S. government has yet been willing to enjoin private enterprise from pursuing the benefits of the research (that would change, should Congress legislate a full ban against therapeutic cloning—a measure being contemplated at this writing). It seems reasonable to expect that those who oppose research on human stem cells would prefer a universal injunction against it. They have found it necessary, however, to settle for a political compromise. It is not a compromise that sits well with many observers. In the words of one British Nobel laureate, "The U.S. is in a real muddle . . . What sort of signal does it send out when the private sector can do anything and the public sector is restricted? How can you take such [policy] seriously?"[27]

It is nevertheless a compromise with teeth. The denial of federal funds is a constraining circumstance, because the federal government could bring formidable resources to bear on the research, were that permissible; because those resources would support the kinds of fundamental research that are not generally performed by private enterprise, yet are exceedingly important to advances with stem cells; and because the exclusive use of private funds threatens to entangle the research in commercial concerns over intellectual property. Given the political influence of abortion opponents, the United States appears destined for a prolonged debate over the extent to which it is going to pursue the potential of human stem cells.

There is some slight chance that science itself may provide a solution to the impasse. It is possible that "politically correct" stem cells might be obtained from adult sources, including blood, bone marrow, muscle, and other tissues. The present uncertainty is whether these forms of stem cells will prove to be as versatile as those derived from early embryos. Many scientists suspect that they will not, and the evidence to date is not promising.

Evolution

There is perhaps no more profound disconnect between the community of science and the general public than the continuing strife over evolution. Properly defined, evolution is the view that all species have a common origin in the remote past of living matter. The term is popularly associated with Charles Darwin, but the idea preceded him. What Darwin did was to amass an immense body of data to support the reality of evolution. These data formed the basis for his monumental book *On the Origin of Species by Means of Natural Selection; or, Preservation of Favored Races in the Struggle for Life*, widely regarded as one of the cultural landmarks in the history of humankind.

But Darwin complicated matters for subsequent generations by using his book not only to argue for evolution itself, but also, as the title of the book foretells, to propose a mechanism by which life became so abundantly diversified from a single source. He called the mechanism "natural selection," although it is more popularly known as "survival of the fittest" (Darwin coined that phrase as well). Darwin was timid (or shrewd) enough to say virtually nothing about humankind in his first description of evolution by natural selection. But in a subsequent book, *The Descent of Man, and Selection in Relation to Sex*, he threw down the gauntlet and argued that Homo sapiens had evolved from some lower form. He even suggested a location for the origin of man —Africa, a stunning anticipation of modern scientific orthodoxy. The suggestion flew in the face of white supremacy and may have helped deter the acceptance of Darwinian thought by early twentieth-century Western Europeans.[28]

Virtually all biologists believe that Darwin established evolution as a fact of life (or of life's history, to be more exact), and modern science has added many new dimensions to the evidence.[29] There is also widespread agreement that evolution is based on natural selection, although room remains for argument about the details. In contrast, the school of thought known as "creationism" accepts neither of Darwin's conclusions, arguing instead that the biblical account of creation accurately describes the origin of life and of all living creatures at one stroke. Other cultures have their own creation myths, but these have not engendered the impassioned opposition to evolution that typifies biblical creationism. Of late, even Pope John Paul II has declared that evolution is "more than just a theory," indeed, that it has "proved true."[30]

It has been more than seventy-five years since the Scopes Trial in Dayton, Tennessee, exposed the speciousness of the creationist arguments against evolution.[31] The trial originated from an initiative by the American Civil Liberties Union to challenge a Tennessee law that prohibited the teaching of evolution in public schools. In an effort at public relations, civic leaders of the town of Dayton (population 1,800 at the time of the trial) persuaded twenty-four-year-old John T. Scopes, general science teacher and part-time football coach, to be prosecuted for teaching evolution—as indeed he had been doing; "nobody could teach biology without teaching evolution" was the way one of his acquaintances put it.[32]

The trial took place in the summer of 1925. The outcome was a foregone conclusion. Scopes readily acknowledged his crime, was found guilty by the jury, and fined $100 (the minimum possible penalty). But during the trial, the principal lawyer for the defense, Clarence Darrow, persuaded William Jennings Bryan to take the witness stand. Bryan was a celebrated lawyer, journalist, orator, perennial candidate for the U.S. presidency, and biblical fundamentalist who was spearheading the prosecution in the trial. The ensuing cross-examination by Darrow exposed the flaws in Bryan's fundamentalist beliefs and undermined the credibility of his opposition to evolutionary theory.

It is notable that Bryan himself argued only against the teaching of evolution as a fact, not against its inclusion in the curriculum of public schools as an unproven theory. The position of latter-day creationists is little different; indeed, is sometimes more extreme. They are always given more than a cursory hearing, and on occasion, they come close to success. The State of Kansas recently suffered great embarrassment when its board of education voted to remove the subject of evolution from its public-school curriculum. To its credit, the voting public of Kansas quickly retaliated at the ballot box. The children of Kansas will be taught the facts of life's history after all.

Creationists typically conflate the genesis of species by evolution with the origin of life and even the origin of the universe. It is a convenient confusion. The evidence for evolution is rock solid (indeed, much of it has been found in rock, in the form of fossils), whereas the explanations for the origins of the universe and of life remain hypothetical. Conflating the three allows the creationist attack on "origins theories" to also challenge the authenticity of evolution. In reality, Darwin wrote virtually nothing about the origin of life itself. His preeminent concern was the origin of species.

Creationists often confound the debate in another manner by treating evolution and natural selection as consubstantial. They are not. Natural selection is the widely accepted explanation for how evolution occurs. But biologists still argue about some of the details of natural selection, and creationists use these disputes inappropriately in efforts to discredit the reality of evolution itself.

The very definition of life now relies in part upon the reality of evolution. Biologists define any system as "living" if it displays two properties: the ability to reproduce independently, and the ability to evolve. Once more: nothing in biology makes sense except in the light of evolution.

The newly available chemical sequence of the human genome provides a stunning account of human evolution, a fossil record in DNA that spans more than a billion years and records the major steps in the emergence of Homo sapiens. Only the most obdurate opponents of evolution could now deny its reality. The human genome sequence has also dramatized the kinship within the human family. The DNA of any two people on the planet is likely to be 99.9 percent identical. These findings have added a powerful biological argument against the fictions of race and bigotry. There is no molecular substance to the belief that one portion of the human family is inherently inferior to another;

indeed, no molecular substance to the existence of race in any biological sense of the term.

Evolution in the Public Schools

In 1989, the State of California hammered out a revised version of its Science Framework, a 190-page set of instructional guidelines that, among other things, specify what should be contained in textbooks used by the public schools. This is no small issue, because whatever the large California (and Texas) marketplace asks of publishers, the rest of the nation is also likely to get. True to form for the Golden State, warfare broke out over the treatment of evolution in textbooks of biology. Biologists saw the chance to eliminate all equivocation from the teaching of evolution, whereas religious fundamentalists mounted a rearguard action designed to keep the door open for creationism.

By and large, the biologists had the upper hand. But in order to get final approval from the conservative California State Board of Education, compromises were struck on several tactical points. In particular, the guidelines were altered to include a statement that, out of religious conviction, some people do not subscribe to evolution; to eliminate a description of how molecular comparisons among the DNAs of many different species had added persuasively to the evidence for evolution; to remove a paragraph that added the weight of the U.S. Supreme Court to the legitimacy of teaching evolution; and to delete a statement that declared evolution a "scientific fact"—it would remain a "theory" in the eyes of California students, as it had been for William Jennings Bryan more than seventy-five years ago.

The compromises were treated as minor concessions by state authorities, as welcome but inadequate improvements ("crumbs from the table") by the opponents of evolution, as pedagogical travesties by the scientific community. One state legislator pronounced that the teaching of evolution as scientific fact is "educationally unsound and morally corrupt," another that "there is more proof in the theory of creation than there is in the theory of evolution." One member of the board of education (and, alas, a tenured member of the faculty at the

University of California, Los Angeles) argued that "to give students a good education, you have to give them both sides of an issue," failing to account for the Constitutional proscription against introducing religious doctrine (creationism) into public education. The front page of the *New York Times* and ABC television news declared a "victory" for the "foes of evolution."

Newly minted as a Nobel laureate and perhaps overly confident of the influence I might thus have gained, I sought to dissuade the state board of education from their misguided intentions. In the correspondence that followed, the president of the board defined evolution as a "scientific theory which explains the 'how' of the origins of the universe, earth and life." How can you argue that evolution is established fact, he asked, when NASA is still launching space probes to gather evidence for the "Big Bang," which he referred to as a "particular evolutionary theory." But the Big Bang is no such thing; it is solely a theory about how the universe might have originated, with no direct bearing on either the origin of life (which occurred billions of years after the reputed Big Bang) or the evolution of species (which began with the origin of life and continues to this day).[33]

This correspondence left me simultaneously disconsolate and irate: disconsolate that any educated individual could harbor such profound misapprehensions, irate that such misapprehensions were informing actions by a principal steward of public education (and in my home state, at that). Surely we could expect him to know that evolution and natural selection are legitimate scientific theories, whereas creationism is a religious conviction. Until texts and teaching present an honest and clear image of evolution, the public confusion will continue. It is a confusion that both bewilders and amuses the educated public of other Western nations, where evolution is no more controversial than the oblate spherical shape of our planet. The widely read British scientific journal *Nature* never misses the opportunity to report the latest antics of U.S. creationists.

Although creationism is clearly theological doctrine, its proponents claim to have evidence that supports the biblical account of creation, derived from a pursuit known as "creation science"—an archetypal oxymoron. The claim is not weathering the test of time very well. In its stead has come "intelligent design theory," a more facile argument. Its proponents acknowledge that species have been long in the making, as suggested by evolutionary theory. But they argue that the fundamentals of biology are "irreducibly complex" and thus could not have originated solely from chance, as envisioned in natural selection. There must have been an intelligent designer behind it all. The argument reflects an unwillingness to concede that, as before in science, what presently seems beyond understanding is likely someday to become clear. There is no material evidence to sustain intelligent design theory. Still, at this writing, the proposed addition of the theory to the curricula of public schools in Ohio is being vigorously promoted before the state board of education.[34]

In reality, creation science and intelligent design theory consist mainly of efforts to discredit the evidence for evolution, rather than to adduce evidence for creation and a supernatural designer—the fossil record is not likely to give up Adam's rib. The U.S. courts have, with some consistency, seen through the deception, ruling that creation science has religious intent and, thus, cannot be compelled into curricula by legislation. But out in the hustings of local education, the creationist assault on the First Amendment to the U.S. Constitution continues.

Academic Mischief

Perhaps the most peculiar strife afflicting science arises from within the halls of academe. It comes as a surprise to many, but there is a school of thought among the intelligentsia that considers science to be fraudulent as a way of knowing. According to this school, the supposed objective truths of science are in reality all "socially constructed fictions," no more than "useful myths." Science itself is "politics by other means." The implication is that if any particular piece of science were redone in a different time and different social context, the outcome might be entirely different. In other words, experimental findings are bogus and tell us only what the observer wants us to hear.

This school of thought is known by various names, including "cultural constructivism," "strong program sociology," and "postmodernism." Anyone with a working knowledge of science, anyone who looks at the natural world with an honest eye, should recognize all of this to be arrant nonsense.[35] The laws of physics come out the same, the earth is truly round, the engines of life have the same construction, the genetic code uses the same alphabet and words, whether these are examined in East or West, in this century or the last, by woman or man. That is the power of science, and it is unmatched by any other, more subjective human endeavor. Scientists are fallible. They can misread, misinterpret, mislead, and otherwise err. But the truths that they pursue can be verified by reproducible experiments.

The credibility of the postmodern view of science was badly tarnished when the physicist Alan Sokal succeeded in having a bogus article published in *Social Text*, one of the leading journals of this dubious discipline, with praise and great satisfaction from the editors (until the hoax was revealed). The title of Sokal's article speaks volumes: "Transgressing the Boundaries: Toward a Transformative Hermeneutics of Quantum Gravity." No self-respecting physicist would pretend to know what that means.[36]

It is difficult to say what fuels the postmodern vendetta against science. One explanation could be its provenance: if subjective forces have any role at all in the conclusions of science, it is in the social sciences, and postmodernism is a child of sociology. The objectivity of science may provide another explanation: no other human pursuit is anchored so fully in observation, experiment, and reproducibility; so scientists can ultimately resolve their disputes beyond question, a privilege granted no other discipline. And envy probably plays its part: within academe, science commands the lion's share of resources, wields great influence, and is capable of deep arrogance; none of these sits well with beleaguered humanists. But the root of it all, I suspect, is ignorance—of both the practice and content of science.

My professional life takes place in an academic medical center, where the efficacy and benefits of science are a daily reality. When I first heard the postmodern view of science some years ago, I dismissed it as merely a careerist strategy for success in parochial corners of academia—the product of either guile or ignorance.[37] So I was deeply disappointed when the rubric of postmodernism was taken up by Vaclav

Havel, the widely admired Czech writer and statesman. Soon after gaining international prominence, Havel began to vigorously publicize what he styled as a newfound disenchantment with science.

Havel attributed his disenchantment to the failure of communism, which he regarded as a scientific effort to explain and control social forces: "The fall of Communism can be regarded as a sign that modern thought—based on the premise that the world is objectively knowable, and that the knowledge so obtained can be absolutely generalized—has come to a final crisis."[38] In this statement and many others like it, Havel displayed a failure to distinguish between the social and natural sciences, indeed, between even political philosophy and science.

Out of these dubious origins came Havel's central argument. "Modern rationalism and modern science . . . now systematically leave [the natural world] behind, deny it, degrade and defame it—and, of course, at the same time, colonize it." Those are angry words, although their meaning is difficult to plumb. The anger fueled Havel's apocalyptic conclusion: "This era [of science and rationalism] has reached the end of its potential, the point beyond which the abyss begins."

The reader might expect that, given its dubious credibility, postmodernism has by now run its course. But the movement continues to invent diverting new interpretations of human endeavor. Among its current preoccupations is personal genius, which postmodernist scholars view as not an exceptional human attribute, but the product of "good marketing or good politicking," "intellectual imperialism," or a "patriarchal imposition on feminine achievements." Writing in the *New York Times*, Edward Rothstein trumped these arguments by asking whether the creators of Bach's Goldberg Variations, Beethoven's Opus 111 Sonata, Mozart's *Nozze di Figaro,* or Bartok's string quartets were simply "ordinary folk granted disproportionate attention by changing fashion," then answered sardonically with "just listen."[39]

A Task Force for Mischief

Against all odds, postmodernism at one point reached into the affairs of the U.S. Congress. This odd episode began with the late George

Brown, a congressman from California who admitted to having his own faith in science shaken by the ruminations from Havel. Congressman Brown was trained as a physicist and was a durable and thoughtful friend of science, known admiringly as the "science congressman." But influenced by Havel, he took up some disquieting themes.

Brown complained of what he called a "knowledge paradox": a parallel rise in fundamental knowledge on the one hand and societal dysfunction on the other. He argued that the two trends should be reciprocal; that as science progresses, the problems of society should diminish. He wondered why science has not contributed more to the achievement of national goals and suggested that we may have to change the ways in which we identify research for funding—in particular, by giving greater authority to Congress and the "consumers" of research.[40]

In 1991, Congressman Brown acted on his newfound doubt and commissioned a Task Force on the Health of Research to address the "knowledge paradox." The members of the task force took the bit between their teeth and produced the following conclusions.[41]

First, Congress should exert greater control over the choices of research to be funded. This was cause for concern, given the congressional tendency to "pork" as opposed to rigorous choice.

Second, research should be addressed more immediately to "current political, economic and societal pressures." Implicit in this suggestion is the assumption that it is possible to determine in advance which research will fulfill a national goal. Few scientists would concede that assumption, and even fewer could justify it.

Third, legislative mandates should be used to determine how research is evaluated. "Programs that are failing to meet stated goals should be terminated." By this criterion, I suppose we should now terminate cancer research because it has so far failed to produce a panacea for the disease.

Fourth, it may be preferable to discard peer review as we now know it, in favor of block grants and funding decisions by "smart managers"—individuals who do "research on research."[42] This recommendation flies in the face of the international opinion that rigorous peer review deserves a lion's share of the credit for U.S. dominance in research.

And fifth, the "users" of knowledge should have a greater role in evaluating research performance. I will grant some merit in this, just as I would also welcome the participation of "users," in preference to that of "smart managers," in funding decisions. In recent years, the NIH at least has begun to include "users" such as patients and their advocates in deliberations over research policy.

These suggestions betray expectations that science cannot meet, a misapprehension of its capabilities. They failed to recognize that the motives of public policy cannot mandate success in science: the progress of science is driven by feasibility—science is the art of the soluble, of the possible, to borrow a phrase from the biologist and Nobel laureate Peter Medawar; we can seldom force nature's hand—usually, it must be tipped for us.[43] The task force also slighted the substantial strategic planning that has guided both fundamental and applied research in the United States over the past fifty years, and the plentiful results that have redounded to the benefit of society. These advances did not come from random walks through the vineyards of research.

The members of the task force clearly displayed a bent for the social rather than the physical and biological sciences—there is a scent of postmodernism here. They wrote a script for mischief. The script soon fell into obscurity through benign neglect, as do most such reports from within the federal government (and those outside it, for that matter). But the concerns it voiced and the remedies it offered should give all scientists pause for thought. To quote Bernard Davis again: "Scientists have not conveyed to the public, or even themselves generally appreciated, the importance of [the] barrier between natural science and social problems. The result is unfillable [sic] expectations. An emphasis on the limits of science not only will help eliminate this source of public disaffection but will place scientists in the unusual public position of exhibiting humility."[44]

Congressman Brown and his task force misplaced much of the blame. Science has long since produced the vaccines required to control many childhood infections in the United States, but our society has so far failed to fully deploy those vaccines. Science has long since sounded the alarm about acid rain and identified its principal origins in automobile emissions, but our society has not found the political will to bridle the internal combustion engine. Science has long since documented the medical risks of addiction to tobacco, yet our federal government still spends large amounts of money subsidizing the tobacco industry and cannot bring itself to regulate that industry. Congressman Brown argued, "We must test the hypotheses that link economic and social benefits directly to research."[45] I regard those hypotheses as by now well proven.

Calculating Needs and Returns

Much of the tension over science on Capitol Hill is generated by conflicting demands for money. How much should we spend on science; how much is needed? I think it is fair to say that no one really knows. No one has yet devised a calculus that science and government can both trust. For the moment, science solicits according to perceived opportunity; government appropriates according to what the traffic will bear. We need to improve on that crude and surely capricious formula.

Consider, for example, the now serious imbalance between funding of the NIH and the National Science Foundation (NSF). The 2002 budget for the NIH exceeded $20 billion, whereas that of NSF was a mere $4.5 billion—not much more than the annual increase scheduled for NIH in fiscal 2002. The discrepancy is readily explained. The NIH is charged with the prevention and cure of disease, and thus easily attracts a large and diverse constituency (including the aging and ailing members of Congress themselves), whereas the NSF funds nonmedical research of immense variety and importance, little of which is readily accessible to the sympathies of either Congress or the public.

The imbalance may appear advantageous to the NIH, but it could easily undermine the institutes' objectives in the long run. To an ever increasing extent, progress in biomedical research is relying on techniques and concepts generated by the nonmedical sciences that the NSF supports, including chemistry, physics, engineering, and com-

DOONESBURY   **by Garry Trudeau**

"Let's Go Back: Priorities in Research" by Garry Trudeau. (*Doonesbury* © 1973 G. B. Trudeau. Reproduced with permission of Universal Press Syndicate. All rights reserved.)

puter science. If these disciplines wilt, so too will the prospects for progress in our understanding of human health and disease.

Part of the difficulty is that we have never adequately calculated the return on our investment in fundamental research. Again, it is not clear that anyone as yet knows how to make the calculation. But the available approximations suggest staggering figures.[46] To cite one example from my own purview, it has been estimated that the vaccine against poliovirus now saves the United States more than a billion dollars every year in costs of health care and lost productivity, whereas the cost of developing, producing, and distributing the vaccine can be reckoned in the mere millions. Similar savings would be realized for health care alone if all working adults in the United States were immunized against influenza. We need to improve on calculations of this sort and deploy them in the design of the federal budget. Until we do, science will continue to solicit at an unnecessary disadvantage and be unduly hostage to the vicissitudes of both politics and the economy.

It is probably unavoidable that pragmatism figures so large in the public funding of research. Senator Tom Harkin, a friend of biomedical research, nevertheless lives by the following mantra: "NIH stands for the National Institutes of Health. It does not stand for the National Institutes of Basic Research."[47] Still, science does have a fundamental value that is beyond calculation. To his great credit, Congressman

Brown knew this value well: "Basic research represents a uniquely human quest to achieve intellectual and spiritual insight and growth through scientific inquiry . . . Particle accelerators, spacecraft, cathedrals, and libraries all are essentially similar. They are settings for cultural experience."[48]

Stephen Hawking has phrased the argument even more grandly: "Ever since the dawn of civilization, people have not been content to see events as unconnected and inexplicable. They have craved an understanding of the underlying order in the world. Today we still yearn to know why we are here and where we came from. Humanity's deepest desire for knowledge is justification enough for our continuing quest."[49] In any event, I firmly believe that it is foolish to declare any knowledge forever useless. Let me dramatize this point with an example from archaeology, not high on the public's ranking of relevance. There is, in fact, much of worldly value to be learned from the excavation of the human past. Jared Diamond has made this point nicely by describing how archaeological studies of the Anasazi settlements in precolonial North America have provided valuable lessons in habitat destruction and resource conservation.[50]

Paying for Science

The strife over funding of research that I have just described arose during a time in the early 1990s when the federal budget was lean. In the interim, the debate has been muted by a reversal of fortune. Federal revenues have burgeoned, and strong Congressional leadership has emerged to direct some of that plenty to an enrichment of research budgets. Congress will soon complete an effort to double the budget of the NIH, a laudable step in the right direction, particularly in its implicit endorsement of long-term budgeting for the research enterprise.

But beyond the funding of NIH lies great uncertainty and concern. In February of 2001, a federal commission reported that the general support of research and development in the United States has slipped to a "crisis" level, and that this support should enjoy the same doubling as the more focused budget of the NIH.[51] But caprice still rules:

the appropriations for research budgets must run an annual gauntlet that begins in the White House and continues in Congress; the national economy fluctuates without reference to societal needs; and both presidential and congressional leadership changes, sometimes precipitously.

Our investment in research reflects a distortion of public values. The United States has traditionally spent less than 2 percent of health care costs on medical research. In contrast, the defense industry has spent as much as 15 percent of its budgets on research and development; pharmaceutical companies, generally more than 10 percent; the aerospace industry, 6 percent; automotive companies, 5 percent; the tire and rubber industry, 3 percent. In 1992, the United States spent approximately $1.9 billion on cancer research, $1.3 billion on AIDS, $730 million on heart disease, and a mere $280 million on diabetes. In contrast, the nation spent in excess of $400 billion on military defense, $170 billion on Fords and Chevies (the sum for Japanese imports is too humiliating to be mentioned), $140 billion on "recreational drugs" (all of them presently illegal), $6.9 billion on subsidies so farmers would not grow crops, $4.2 billion on antiballistic missile research (known colloquially as "Star Wars research"), and $1.8 billion on the Nintendo computer game—a dead heat with cancer research.

The absolute numbers have grown during the ensuing years, but otherwise, the picture has not changed substantially—although Nintendo and Star Wars research fell on hard times (the latter is enjoying a revival with the election of George W. Bush to the U.S. presidency). Surely there is room here for fundamental research of all sorts, even for undertakings as ambitious as the Human Genome Project and the Superconducting Super Collider. I made certain to mention these two megaliths because they exemplify a special problem in the funding of science: the threat that we may rob Peter to pay Paul, small science to pay large. With costs running into billions of dollars, the two projects represent the sort of gigantism that is anathema to many scientists. Freeman Dyson finds the hostility misplaced: "We cannot calculate from general principles the optimal size of a scientific project, any more than we can calculate the optimal size of a whale."[52]

Dyson happened to like the Human Genome Project because of

BIG SCIENCE   LITTLE SCIENCE

"Big Science/Little Science." © 2002 Sidney Harris. (Reproduced by permission of the artist.)

its strategic pragmatism. His affection was well placed: the project is now virtually complete, it has cost less than anticipated, and it will have incalculable value. The Human Genome Project was aided by an unexpected development. The work began as an exclusively federal initiative. But a parallel and vigorously competitive private effort then sprang up, brought technological innovation to the race, and helped drive the work to early fruition. Still, social strife emerged, in the form of acrimonious disputes over access to the data and what commercial benefit the private effort should be permitted.

Dyson was less enthusiastic about the Super Collider, urging that we build "several clever accelerators instead of one dumb accelerator."[53] It goes without saying that the clever accelerators will require new ideas before they can be built, and it is assumed that they will be less expensive than the dumb one. The Super Collider failed to gain congressional support and never became more than an expensive and unoccupied tunnel. But this much should be said: the device would have cost less than a single Sea Wolf submarine, which we have not hesitated to build in multiples. The Nobel laureate physicist Steven Weinberg believes that we should have built the collider, because without it "we may not be able to continue with the great intellectual adventure of discovering the final laws of nature."[54]

As for the International Space Station, that champion Gargantua of the moment, suffice it to say that its proponents have been using the specious prospect of biomedical experiments in space as a major justification for the behemoth. That argument will not bear scrutiny. No biomedical scientist of my acquaintance believes there is anything we could do on the space station that would come close to justifying its price tag of more than $30 billion.

We must move beyond categorical debates over big and small science to distinctions made on other criteria, such as intellectual promise and quality, potential utility, and intelligence of design. As Steven Weinberg writes: "Arguing about big science versus small science is a good way to avoid thinking about the value of individual projects."[55] Big science is not inherently bad. But it must be judged with meticulous care and honesty.

We need an objective standard by which to design the funding of research. Take biomedical research as an example. Why not follow the lead of the pharmaceutical industry? Why not budget 10 percent of the total cost of health care for biomedical research? Why not obtain at least a portion of the budget for research with levies on the health care industry, which is nourished by research, yet remains one of the few industries that expends virtually none of its revenues on research?[56] Why not build this principle into the national health plan that sooner or later our federal government seems destined to create?

It is not easy to raise claims for science in the face of the social ills and external threats that beleaguer our nation, problems that can make the limits on science seem a parochial issue. But our nation remains prosperous and generously endowed. If in a fiscal instant we can find the enormous sums required to rescue ailing industries and foreign economies, wage war against Iraq in one decade and the Taliban in another, and implement a $1.7 trillion tax reduction, surely we can find the resources to secure an enlightened future, for ourselves and for the generations that follow.[57]

The federal support of fundamental research must be sustained. The community of science should not be reticent. We speak in self-interest, of course, as do all beneficiaries of the federal patron. But we also speak for one of the noblest endeavors of humankind. If we do not seize the day, the politics of greed may foreclose on the future.

Disappointment

Disenchantment with science arises in part from unrealistic expectations. So much has been achieved that far more is expected than we can hope to deliver. Why has malaria not been eradicated by now? Why is there still no cure for cancer and AIDS? Why is there not a more effective vaccine for influenza? When will we cure the common cold? When will we be able to produce energy without waste products? The litany of disappointment seems infinite.

This disappointment is in part the fault of scientists themselves. The fault lies with "scientism"—the belief that the methods of the natural sciences are the only means for obtaining knowledge and understanding. We would do well not to claim science as the exclusive source of truth about human existence—this despite the distinguished philosopher Saul Kripke, who once lamented that philosophers had become preoccupied with the search for meaning and had abandoned to scientists the search for truth. "What scientists need to avoid," wrote the philosopher Mary Midgley, "is fundamentalism—the conviction that the particular imaginative vision espoused by their own party of current scientists is a solitary gospel which must always prevail."[58]

Yet the hyperbole of scientists often betrays a strain of scientism. Stephen Hawking's best-seller *A Brief History of Time* concludes with the now famous line that physicists may someday "know the mind of God" (poetic license, I am sure, but arresting to the innocent).[59] In reporting new evidence for the cosmic Big Bang, scientists hinted at the sighting of a godly hand, which attracted more attention in the press than the exciting (but far from decisive) science itself. The redoubtable

Francis Crick subtitled a book *The Scientific Search for the Soul,* even though it was mainly about how we see.[60] The sociologist Dorothy Nelkin complains that the gene has become a cultural icon, a means by which to explain falsely all of human behavior and fate.[61] While serving as the first director of the Human Genome Project, James Watson was fond of justifying that project with words that dramatized Nelkin's complaint: "We used to believe that our fate was in the stars; now we know that it is in our genes."[62] And Nobel laureate Walter Gilbert opined that once we have the complete sequence of the human genome, "we will know what it is to be human."[63] Well, we have the sequence now (or most of it, at least), and what we know principally is that we are stunningly similar to chimpanzees in the makeup of our genes.[64]

Richard Feynman offered a corrective to all this:

> Which end is nearer to God . . .? Beauty and hope, or the fundamental laws? I think that the right way, of course, is to say [that] not just the sciences but all the efforts of intellectual kinds, are an endeavor to see the connections of hierarchies, to connect beauty to history, to connect history to man's psychology, man's psychology to the working of the brain, the brain to the neural impulse, the neural impulse to the chemistry, and so forth, up and down, both ways . . . And I do not think that either end is nearer to God. To stand at either end, and to walk off that end of the pier only, hoping that out in that direction is the complete understanding, is a mistake. And to stand with evil and beauty and hope, or to stand with the fundamental laws, hoping that way to get a deep understanding of the whole world, with that aspect alone, is a mistake. It is not sensible for the ones who specialize at one end, and the ones who specialize at the other end, to have such disregard for each other.[65]

Distrust

Disappointment transforms easily into distrust. That transformation has been embodied by playwright and AIDS activist Larry Kramer, who spoke for many sufferers of AIDS when he complained that science has yet to produce a remedy for the disease. Kramer placed much of the blame on the NIH, which he denigrated as "a research system that by law demands compromise, rewards mediocrity and actually punishes initiative and originality."[66]

I cannot imagine what law Kramer had in mind, and I cannot agree with his description of what the NIH expects from its sponsored research. I have assisted the NIH with peer review for more than twenty-five years. The standards used have always been the same, seeking work of the highest originality, but demanding rigor as well (a demand that some may find frustrating, but that cannot be compromised—there is too much at stake).

There are critics like Kramer (some from within the house of science, I regret to say) who seek to replace peer review of research with a less formal and more agile system of their own—recall the "smart managers" of the Brown task force. They are wrong. First, because such systems are too easily corruptible. And second, because the approach we have now works well, whatever its blemishes. Revision may be in order, but certainly not rejection.

The proof is in the pudding. Biomedical research in the United States has unearthed usable knowledge at a remarkable rate, bringing us international leadership in the battle against disease and the search for understanding, and earning us the admiration of other nations throughout the world. It is most unlikely that we could have achieved all of this if we did business the way Kramer and critics like him claim.

The bitter outcry from AIDS activists was echoed in the 1992 film *Lorenzo's Oil.* The film tells the story of Lorenzo Odone, a child who suffered from a rare hereditary disease known as adrenoleukodystrophy (ALD). The disease destroys the insulation of nerve fibers, cripples many neurological functions, and leads slowly and erratically to death.

Offered no hope by the attending physicians, Lorenzo's desperate parents scoured the medical literature and turned up a possible remedy: administration of two natural oils known as erucic acid and oleic acid. In the face of skepticism from specialists, Lorenzo was given the oils and, in the estimation of his parents, ceased to decline, perhaps even improved marginally. It was a courageous, determined, and even

reasoned effort by the parents. Whether it was effective is another matter. The course of Lorenzo's disease proved to be little different from
that of many other children with the same affliction.

The film portrayed the treatment of Lorenzo as a success, with the
heroic parents triumphant over the obstructionism of medical scientists. It ended with a montage of parents testifying that the oils had
since been used successfully to treat ALD in their own children. Absent were the parents who had tried the oils with bitter disappointment. And all of this is only anecdotal information. Properly controlled studies are still in progress. To date, these have not given much
cause for hope. At the time the film was made, more than one hundred
children with ALD had received the oils in controlled studies, without
showing any substantive improvement. The erratic course of the disease can lead to cruel illusions. But death remains the inevitable outcome.

Lorenzo's Oil is deeply troubling in its portrayal of medical scientists
as insensitive, close-minded, and self-serving; and in its impatience
with rigorous clinical trials as needlessly wasteful of time—echoing an
early outcry from some AIDS activists.[67] Paradoxically, the film seems
to endorse the legitimacy of science. Lorenzo's parents turned to obscure research literature and biochemical reasoning to find their remedy. (Lorenzo's father has since received an honorary degree from at
least one university.) The villain of the story is not science itself but
scientists themselves, seen through the eyes of two despairing and intelligent human beings. One line spoken by Lorenzo's father late in
the film encapsulates the argument: "These scientists have their own
agenda and it is different from ours." Here is a complaint that physicians and scientists cannot take lightly. It accuses scientists of placing
their personal advancement over the commonweal, of preferring eso-
teric inquiry over the application of science to practical ends, of re-
sentment against the uninitiates who come to science with sound
questions and ideas. The accusations ring uncomfortably true.[68]

As if on cue, isolation of the damaged gene responsible for ALD was
reported soon after *Lorenzo's Oil* had completed its screenings in the
United States. Thus, the exact biochemical defect responsible for the
disease is known at last. Its identity could not have been predicted

from what was known before. The stage is set for the development of decisive clinical testing and therapy, although therapy may still be long in coming.[69]

Lorenzo's Oil reflects three "myths" identified by the bioethicist Arthur Caplan. First, that "cures can be found if only bureaucracy and red tape will get out of the way"; second, that "perseverance, hard work and love can conquer any ailment"; and third, that "mainstream science is indifferent" to the suffering of patients and their families, choosing instead any course that will advance a career.[70] These myths lead easily to disenchantment with scientists. But they also reflect a faith in science itself that can create common cause between the activists who seek more rapid progress against human disease and the scientists who are in a position to make that progress.

Suspicion

While some distrust the motives of scientists, others distrust their practices. During 1989 and 1990, a subcommittee of the U.S. Congress spent an extended period and considerable funds to investigate whether Professor David Baltimore had been party to fraud in work that he and others had published in a major scientific journal. Professor Baltimore was on the faculty of the Massachusetts Institute of Technology when the work was performed, but became president of Rockefeller University during the time of the investigation. It was an affair guaranteed to attract great attention and inflict great harm.

In truth, the investigation never formally charged Baltimore himself with fraud (although members of the congressional staff made such accusations to the press). The accusation was made against his principal collaborator, Tereza Imanishi-Kari, who had worked independently in her own laboratory. The charges were brought by a young scientist named Margaret O'Toole, who had worked with Imanishi-Kari and who pressed the case with great determination. Baltimore himself was deemed by some to have been inadequately vigilant about the veracity of the data that Imanishi-Kari produced, and for this, he was hauled before a congressional subcommittee and subjected to an extensive investigation.

David Baltimore is a Nobel laureate and among the most distinguished of contemporary biomedical scientists—a man of almost preternatural talents. The congressional staff treated the investigation as something akin to a big game hunt, whose object was to humiliate a renowned individual for refusing to capitulate to dubious charges. Representative John Dingell, chair of the Congressional subcommittee conducting the investigation, is reported to have said: "I am going to get that son of a bitch. I am going to get him and string him up high."[71]

The resources deployed in the investigation were both intimidating and ludicrous. They featured an aggressive congressional staff with seemingly limitless powers and resources for investigation; a practiced whistle-blower, borrowed for consultation from the staff of the NIH, even though he had no semblance of expertise in the research at issue; even agents of the U.S. Secret Service, who spent many months and many more taxpayers' dollars examining subpoenaed laboratory notebooks for evidence of falsification.

The U.S. attorney in Baltimore, Maryland, had a look at the evidence and declined to prosecute. In the end, an appeals board assembled by the U.S. Department of Health and Human Services found the case against Imanishi-Kari to be without merit. As for Baltimore, his only offense was to have challenged the credibility of a powerful U.S. congressman and his staff. But by the time exoneration came, grave damage had been done. Baltimore had been obliged to resign as president of Rockefeller University (he later became president of the California Institute of Technology—California's gain, New York's loss), and Imanishi-Kari spent years in professional limbo, a diminution from which she may never fully recover (although she regained her academic position and research support).

Is Congress the venue, is congressional investigation the manner in which the veracity of research and the misconduct of scientists should be explored? Much of what troubled the congressional investigators were in reality practices that reflect the ethos of science—the robust counterpoise of success and failure, of error and correction, of mutual trust and lively criticism, by which science proceeds; there is little understanding of these in the average congressional mind. If U.S. science were shot through with corruption, as some of its critics in Congress seemed to believe, how could it have achieved or maintained the dizzying pace of discovery that has characterized the recent decades in research? Each of us in science is utterly dependent upon the truthfulness of our colleagues. The success of science is built on integrity, and that success has never been greater than in our age.

Disdain

If science bewilders and disappoints some, it repels others. A few years ago, Alan Bloom's book *The Closing of the American Mind* appeared on coffee tables and best-seller lists around the land. The book found a sympathetic readership even among some academics. To my eye, however, it was primarily a tedious effort to blame rock music on Nietzsche and Kant. I agree that someone needs to take the blame for rock music, but Nietzsche and Kant will not do.

In his book, Bloom likened science to "the absurdity of a grown man who spends his time thinking about gnats' anuses." "We have been too persuaded of the utility of science," Bloom ranted, "[to perceive] how shocking and petty the scientist's interests appear . . . If science is just for curiosity's sake, which is what theoretical men believe, it is nonsense, and immoral nonsense, from the viewpoint of practical men."[72]

Despite the vitriol, it would be awkward to suggest that these were the ravings of a deranged fanatic. Alan Bloom was a distinguished professor at the University of Chicago and his book carried an admiring introduction from Nobel laureate Saul Bellow. Bloom owed more than he might have realized to Nietzsche, who described university teaching and research as "[a] molish business, the full cheek pouches and blind eyes, the delight at having caught a worm, an indifference towards the true and urgent problems of life"—once again, a preference for practical over theoretical men.[73] These sentiments reverberate through the angry skepticism of *Lorenzo's Oil*, the distrust of Larry Kramer, the disdain of Professor Bloom, the disquiet of Congressman Brown.

Bloom is now deceased, so I belittle him with some reluctance.[74] On the other hand, he did intend his words and reputation for posterity, so he remains at risk posthumously. I am reminded of Pedro Guerrero.

The former Los Angeles Dodger and St. Louis Cardinal once complained that he is misunderstood by the public because "newspapers write what I say, not what I mean."[75] Could it be that Professor Bloom did not mean what he wrote? Would he have wittingly demeaned the great quests of natural science, such as the search for a Grand Unifying Theory of matter, the exploration of our origins in evolution and of how the brain engenders mind, or the explication of how a single cell becomes the glory of the human organism? Those who find no philosophy here, no poetry, no human perspective, are in my view either ignorant or insensate. (I am not alone in my dismay over Professor Bloom. Jill Kerr Conway, former president of Smith College, has described him as "just another aging male misogynist, much less gifted than Tercullian or any of a long chain of predecessors." But that reflects another realm of offense.)[76]

Alas, Alan Bloom has company, the poet John Ciardi among them:

> To the laboratory then I went. What little
> right men they were exactly! Magicians
> of the microsecond precisely wired
> to what they cared to ask no questions of
> but such as their computers clicked and hummed.
>
> It was white-smocked, glass, and lighted Hell.
> And their St. Particle the Septic sat
> lost in his horn-rimmed thoughts. A gentlest pose.
> But in the frame of one lens as I passed
> I saw an ogre's eye leap from his face.[77]

From whence this ogre? What is it the humanist fears in science and finds so repugnant? Science offers what Lewis Thomas once called "the best way to learn how the world works." Or is that the problem—an understanding too plain, too clear to be further reified by poetry? Wordsworth comes to the rescue here: "Poetry is the impassioned expression which is on the countenance of all science . . . The remotest discoveries of the Chemist, the Botanist, or Mineralogist, will be as proper objects of the Poet's art as any upon which it can be employed, if the time should ever come when these things shall be familiar to us."[78] Note Wordsworth's recognition that poets would first have to understand the doings of science before their art would apply. It was also Wordsworth, however, who wrote: "Our meddling intellect / Misshapes the beauteous forms of things; / —We murder to dissect," and he has generally been regarded as an opponent of science.[79] I remain uncertain as to where he really stood—I sense ambivalence.

A postmodernist poet of my acquaintance complains that it is in the nature of science to break things apart, thereby destroying the "mysterious whole." That view ignores the new wonders that unfold when the mystery is solved, revealing the intricacies by which the natural world achieves its ends.

Ignorance

Fear, distrust, and disdain are each in their own way impediments to science. But they all stem from ignorance, and ignorance is our deepest malady. No one has written better of this than the American literary critic and novelist Lionel Trilling: "Science in our day lies beyond the intellectual grasp of most men [Trilling chose not to cast aspersions on women]. This exclusion of most of us from the mode of thought which is habitually said to be the characteristic achievement of the modern age is . . . a wound given to our intellectual self-esteem," creating "a diminution of national possibility, . . . a lessening of the social hope."[80]

Trilling wrote these words many years ago; they are even more apposite now. The problem is before us daily in the United States: in the evidence of woeful scientific literacy among our public; in the failures of our elementary and secondary schools to teach science well (if at all); in the rancorous disputes over the place of science in the general curricula of our undergraduate colleges (far too often, an introductory course in psychology suffices—or nothing at all is expected); in the bewilderment of laborers, accountants, lawyers, poets, politicians, even physicians, when they confront the material of science. The consequences are dire.

Our high school students display appalling inadequacies when tested in physics, chemistry, biology, or math.[81] In a recent Gallup Poll, only 12 percent of U.S. respondents acknowledged evolution as the explanation for the origin of the human species, an explanation that biologists consider beyond doubt (45 percent endorsed the biblical explanation, 37 percent subscribed to "intelligent design"). Many do not know that the earth circles the sun. In a committee hearing some years ago, a prominent member of Congress made national news by confusing the prostate gland with the testes. A former president of the United States was reputed to consult astrological predictions before making decisions of state.

But do the practitioners of science even understand one another?

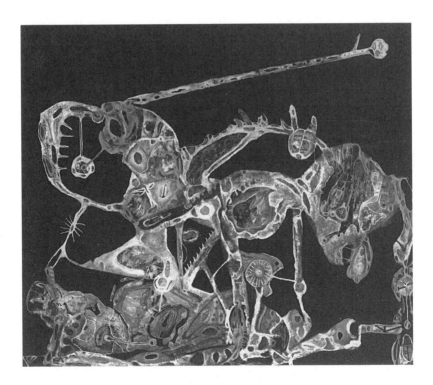

Don Quixote by Charles Seliger, 1944. (© Charles Seliger; reproduced by permission of the Whitney Museum of American Art, New York City, and the Michael Rosenfeld Gallery, New York, N.Y.)

Some years ago, the media reported a Russian satellite that gathers solar light to illuminate large areas in Siberia. "They are taking away the night," I thought; "they are taking away the last moments of mystery. Is nothing sacred?" "Are we crazy?" asked Bill McKibben of the scheme, writing in the *New York Times*. Then again, what do physicists think of biologists' effort to decipher the human genome (now virtually completed), and to recraft it, ostensibly for the better (voluntarily embargoed for the moment)?

I once wrote an article about cancer genes for *Scientific American*. I took great pains to make the text readily accessible: I consulted students, journalists, laity of many stripes. When these consultants had all approved, I sent the manuscript to an acquaintance who is a solid-state physicist of considerable merit. One week later, the manuscript came back, with a message: "I have read your paper and shown it around the staff here. No one understands much of it. What exactly is a gene?"

Robert Hazen and James Trefil have many such anecdotes, which they use to dramatize their advocacy of general science education: twenty-three geophysicists who could not distinguish between DNA and RNA; a Nobel Prize–winning chemist who had never heard of plate tectonics; biologists who thought that string theory had something to do with pasta.[82] We may be amused by these circumstances, but we should also be troubled. If science itself is no longer a common culture, what part of that culture can we expect the laity to grasp? We are all afflicted by what Robert Oppenheimer called "a thinning of common knowledge."[83] We should be seeking a remedy together.

Science and the Classroom

How has all of this come to pass? Lionel Trilling knew the problem in his time: "No successful method of instruction has been found . . . which could give a comprehension of science to those students who are not professionally committed to its mastery and especially endowed to achieve it."[84] And there the problem lies today, perplexing to our educators, ignored by all but the most public-minded of scientists, bewildering and vaguely disquieting to the general public.

Simply put, we have thoroughly botched the job of teaching science.

In many elementary schools, we hardly do it all. In many secondary schools, the curricular materials and strategies are inadequate, the teachers poorly prepared and demoralized. Worse yet, the sequence of teaching science ignores the realities of nature. Typically, we teach biology first because most students find it the easiest of sciences to follow and to enjoy. Chemistry and physics come later. This sequence creates frightful consequences, such as students memorizing the amino acids found in proteins without the slightest sense of what an acid might be, let alone an amino group. In such circumstances, there can be no mystery about why more of our youth do not find science attractive.[85] We should either teach physics first, followed by chemistry and biology; or better yet, integrate the three into a coherent picture of how the world is built and run.

In undergraduate colleges, we remain thoroughly confused about who should study science and how much they should study. The distinguished Harvard chemist Frank Westheimer spent years arguing that science plays too small a role in general education, placing the blame on both recalcitrant humanists who do not want to yield academic turf, and science faculty who do not wish to deal with the "unwashed."[86] A glance across the collegiate landscape today suggests that either few were listening, or few agreed, or few cared enough to do anything.

Westheimer himself never carried the day at Harvard, and he had harsh words about the place: "The vast majority of students who graduate from Harvard are, in a real sense, uneducated because they know almost no science."[87] He was lamenting the architecture of a core curriculum approved by the Harvard faculty in May of 1978. That curriculum remains in place today, allowing Harvard students to graduate with only one-sixteenth of their course work in the sciences.[88] Westheimer may have been offering too strong a tonic, arguing that there should be only one level of collegiate instruction in science, a level that would give no quarter to the students without ambitions in science. "If scientists try to teach nonscientists molecular biology without chemistry, or quantum theory without mathematics, they are unlikely to succeed."[89] There is little realism in that statement: we need a new definition of success.

In the face of these great problems, our nation has allowed the means of primary and secondary education to deteriorate. In doing so, we have incurred great risk, described seventy years ago by the philosopher Alfred North Whitehead:

> The art of education is never easy. To surmount its difficulties, especially those of elementary education, is a task worthy of the highest genius . . . [But] when one considers . . . the importance of this question of the education of a nation's young, the broken lives, the defeated hopes, the national failures, which result from the frivolous inertia with which it is treated, it is difficult to restrain within oneself a savage rage. In the conditions of modern life the rule is absolute, . . . [a country] that does not value trained intelligence is doomed.[90]

We have not heeded Whitehead's warning and it has retained all of its original prescience. Our elementary and secondary teachers are neglected, disrespected, inadequately compensated, and improperly prepared. Many of our children attempt to study in the midst of physical squalor and personal decay. We can expect little improvement in how our youth learn until we have changed all of that. The change will require great resolve: we have allowed the deterioration to run very deep.

Soon after the announcement of my Nobel Prize, I was asked to visit Lowell High School in San Francisco. The quality of the student body, the sophistication of instruction, and the intensity of study at this school are nationally renowned. You would expect an exceptional place in every regard. When I arrived for my visit, I was met outside the front door by a delegation of students—a gesture that struck me as unnecessary for the arrival of a mere adult. I soon understood their purpose: they had come to apologize in advance for the deplorable state of the halls within, embarrassed by something for which they were not responsible. In that moment on the front steps, I felt indicted of grave neglect as a parent, as a citizen and taxpayer, and as an educator. I cannot repeal the indictment; none of us can. We simply must do better.

The teaching of science has been a principal victim of the decline in public education. Those who do not choose science as a career are largely ignorant of its ways, its achievements, and its limitations. They are not prepared to think critically about how science should be used. So we are now at risk of the fate predicted by Henry Adams in 1862: "Man has mounted science, and is now run away with. I firmly believe that before many centuries more, science will be the master of man. The engines he will have invented will be beyond his strength to control."[91] We of science can no longer leave this problem for others to solve. Indeed, it has always been ours to solve, and all of society is paying for our neglect in precious coin.

Coda

The enterprise of science embodies a great adventure: the quest for understanding in a universe that may be "infinite in all directions, not only above us in the large but also below us in the small";[92] the quest for understanding on behalf of life, whose great gift to our planet is diversity, but which remains "a little glow, scarcely kindled yet, in these void immensities."[93]

We have begun the quest well, by building a science of ever increasing power, a method that can illuminate all that is living. Consequently, the community of science is admired, but also feared, distrusted, even despised. It offers hope for the future, but also moral conflict and ambiguous choice. The difficulties of going forward will be large, but they pale in comparison to what we would never gain by holding back.

Notes

Preface

1. Francis Crick, *What Mad Pursuit* (New York: Basic Books, 1988), p. 83. The uppercase letters and italics are from the original.
2. Marjorie Garber, *Academic Instincts* (Princeton, N.J.: Princeton University Press, 2001), p. 39.
3. The aphorism on successful men is widely used, but its origin is not known to me. Success is not necessarily an unblemished blessing. Ambrose Bierce defined success as the "one unpardonable sin against one's fellows": *The Devil's Dictionary* (New York: Dover, 1993), p. 122.

1. The Phone Call

1. The account of Alfred Nobel and his prizes is based on information in Ragnar Sohlman and Henrik Schuck, *Nobel: Dynamite and Peace* (New York: Cosmopolitan Book Corporation, 1929); Thomas Hellberg and Lars Magnus Jansson, *Alfred Nobel* (Stockholm: Alno Production KB, 1983); Ragnar Sohlman, *The Legacy of Alfred Nobel* (London: The Bodley Head, 1983); Kenne Fant, *Alfred Nobel: A Biography* (New York: Arcade, 1993); and Burton Feldman, *The Nobel Prize* (New York: Arcade, 2000).
2. Fant, *Alfred Nobel*, p. 20.
3. Ibid., p. 39.
4. The genesis of the Nobel family's interest in nitroglycerine has been told in various ways. Some describe a direct contact between Alfred and Sobrero (ibid., p. 96), others do not (Feldman, *Nobel Prize*, pp. 28–30). But all sources agree on the essential role that Alfred played in the introduction of nitroglycerine to the Nobel product line, and in its perfection into a safe and practical explosive. Alfred's success prompted protests from Sobrero, who felt that his priority in the discovery of nitroglycerine had been neglected. Nobel avoided any public dispute with Sobrero and was painstakingly courteous with him in private correspondence (see ibid., p. 98).

5. Hellberg and Jansson, *Nobel,* p. 103. Antonio Sobrero also felt pangs of conscience at what he had unleashed on the world, saying that he was "almost ashamed to admit" to being the discoverer of nitroglycerine (Fant, *Alfred Nobel,* p. 97).

6. As quoted in Sohlman, *Legacy,* p. 9.

7. As quoted in ibid., p. 54.

8. As quoted in Fant, *Alfred Nobel,* p. 267. Despite the general influence of von Suttner, it was Nobel himself who first conceived his prize for peace (pp. 270–271).

9. Hellberg and Jansson, *Nobel,* p. 76.

10. Fant, *Alfred Nobel,* p. 177.

11. Hellberg and Jansson, *Nobel,* p. 139.

12. Michael A. Bernstein, "The Faux Nobel Prize," *San Diego Union Tribune,* October 13, 2000, p. B11.

13. Sylvia Nasar, "The Sometimes Dismal Nobel Prize," *New York Times,* October 13, 2001, p. C3.

14. Feldman, *Nobel Prize,* p. 353.

15. As of 2001, the value of the Nobel Prize in each category is approximately $1 million. That sum is divided among the recipients in each category, sometimes equally, sometimes in unequal proportions specified by the Nobel committee.

16. Niels Bohr ranked among the premier physicists of the twentieth century. At the time of the Nazi raid, he was already on his way to Los Alamos, New Mexico, to assist in production of the atomic bomb. The fate of the Nobel medals in Neils Bohr's institute was originally reported by George de Hevesy, *Adventures in Radioisotope Research* (London: Pergamon Press, 1962), p. 27. A more accessible account is in David Bodanis, $E = mc^2$: *A Biography of the World's Most Famous Equation* (New York: Walker and Company, 2000), p. 153.

17. An extended personal account of a "Nobel Week" in Stockholm is in Gerald Weissman, *The Woods Hole Cantata: Essays on Science and Society* (New York: Houghton Mifflin, 1986), pp. 193–210.

18. Sohlman, *Legacy,* pp. 132–133.

19. Ibid., p. 132.

20. Sohlman and Schuck, *Nobel,* p. 1.

21. As quoted in ibid., p. 243.

22. For details, see Kant, *Alfred Nobel,* p. 283.

23. Hellberg and Jansson, *Nobel,* p. 138.

24. For an authoritative account of the deliberations that eventually conferred the Nobel Prize on Albert Einstein, see Abraham Pais, *Subtle Is the Lord* (Oxford, Eng.: Oxford University Press, 1983), pp. 502–511.

25. Ibid., p. 503.

26. As quoted by Richard Stone, "At 100, Alfred Nobel's Legacy Retains Its Luster," *Science* 294 (2001): 288–291. The quotation is on p. 291.

27. The life and work of Marie Curie have been recounted by Rosalynd Pflaum, *Grand Obsession: Madame Curie and Her World* (New York: Doubleday, 1989).

28. All told, twenty-one women have received the Nobel Prize to date.

29. The three repeaters besides Marie Curie are John Bardeen, two prizes in physics; Frederick Sanger, two prizes in chemistry; and Linus Pauling, one prize in chemistry, the other in peace.

30. As quoted in Feldman, *Nobel Prize*, p. 8.

31. This thought is treated more expansively in ibid., pp. 4–9.

32. Anne Sayre, *Rosalind Franklin and DNA* (New York: W. W. Norton, 1975).

33. Max F. Perutz, "Discoverers of Penicillin," in *Is Science Necessary? Essays on Science and Scientists* (New York: E. P. Dutton, 1989), pp. 149–163. The quotation is on pp. 162–163.

34. Hellberg and Jansson, *Nobel*, p. 139.

35. Ibid.

36. Nasar, "Somewhat Dismal Nobel Prize."

37. Sylvia Nasar, *A Beautiful Mind* (New York: Simon and Schuster, 1998).

38. For more about the personal consequences of the Nobel Prize, see Harriet Zuckerman, *Scientific Elite* (New York: Free Press, 1977), and Feldman, *Nobel Prize.*

39. The material on Subramanyan Chandrasekhar is taken from Kameshwar C. Wall, *Chandra* (Chicago: University of Chicago Press, 1991). See especially the final section, entitled "Conversations with Chandra," pp. 245–307. The quotations are from pp. 296–298.

40. See Feldman, *Nobel Prize*, p. 116.

41. Jared Diamond, "A Tale of Two Reputations," *Natural History* 110 (2001):20–24.

42. James D. Watson, *The Double Helix* (New York: Atheneum, 1968).

43. Francis Crick, *What Mad Pursuit* (New York: Basic Books, 1988), p. 81.

44. The point about competition can be made with examples from the remote past: in the seventeenth century, Isaac Newton and Gottfried Leibniz independently discovered the calculus, precipitating one of the most vitriolic disputes over priority in the history of science; and in the nineteenth century, Charles Darwin was moved to complete his magnum opus on evolution and its mechanism only after learning that Alfred Russell Wallace had come upon the same ideas while suffering from malaria in Malaysia—Wallace had formulated the theory of natural selection (survival of the fittest) pretty much the same as had Darwin. Wallace

proved to be uncommonly generous, giving his own book on evolution the title *Darwinism.*

45. Van Wyck Brooks, as quoted by Alfred Kazin, *Writing Was Everything* (Cambridge: Harvard University Press, 1995), p. 151.

46. William Butler Yeats, "The Choice," in *The Collected Poems of W. B. Yeats* (London: Papermac, 1982), p. 278.

2. Accidental Scientist

1. I thank Freeman Dyson, who first urged me to write the personal story contained in this chapter. Portions of the text were adapted from an autobiographical sketch published by the Nobel Foundation in Tore Frangsmyr, ed., *Les Prix Nobel* (Stockholm: Almqvist and Wiksell International, 1989), pp. 215–219.

2. In 1881, the physicist Albert Michelson, assisted by the chemist Edward Morley (who was also a trained theologian), performed a series of experiments on the speed of light. Their results refuted a then-popular idea that light was propagated by undulations of a hypothetical material known as "ether" (no relation to the homonymous anesthesia). This finding proved to be a harbinger of Einstein's theory of relativity. We have since learned that light is propagated in a particulate form known as photons, which can also behave as waves. In 1907, Michelson became the first American scientist to receive a Nobel Prize.

3. The experience of students at liberal arts colleges can be a powerful lure into the life of the mind. For example, in the United States, small colleges produce twice as many Ph.D. scientists per graduate as do baccalaureate institutions in general and even hold their own with the record of the premier research-intensive universities. The intellectual and social intimacy between faculty and students at small colleges probably accounts for much of the attractive force. At the time, I found the life and responsibilities of my college faculty to be little short of idyllic. Of course, I was ignorant of salary scales and the unfortunate struggle for self-esteem on the part of many Ph.D.'s who spend their lives teaching—Clark Kerr, a former president of the University of California, has suggested that faculty at state colleges (as opposed to those at the generally more prestigious and research-intensive state universities) often view their institutions as "graveyards of disappointed expectations"—see his book, *The Gold and the Blue,* vol. 1 (Berkeley: University of California Press, 2001), p. 174. It should not be so. See also Thomas R. Cech, "Science at Liberal Arts Colleges: A Better Education?" *Daedulus* (Winter 1999): 195–216.

4. I recall being tempted by the career of historian, but dismissing that because

it did not seem sufficiently altruistic. Now I read history at every opportunity and am very grateful to those who write it well.

5. The neurologist and best-selling author Oliver Sacks has written similarly about museums and books. See his autobiography, *Uncle Tungsten* (New York: Knopf, 2001), pp. 57–59.

6. Christopher Jencks and David Riesman, *The Academic Revolution* (Garden City, N.Y.: Doubleday, 1968), p. 206.

7. Francis Crick, *What Mad Pursuit* (New York: Basic Books, 1988), p. 17.

8. The story of how Peyton Rous discovered this virus and the impact of the discovery on cancer research is told in Chapter 4.

9. For a biography of David Baltimore and the story of reverse transcriptase, see Shane Crotty, *Ahead of the Curve* (Berkeley: University of California Press, 2001).

10. Roger Lipsey, *An Art of Our Own: The Spiritual in Twentieth-Century Art* (Boston: Shambhala, 1997), p. 440.

11. Dennis Danielson, "Scientist's Birthright," *Nature* 410 (2001): 1031.

12. The manner in which artists and musicians have progressively broken more of their own rules has been explored by Leonard B. Meyer, *Music, the Arts, and Ideas* (Chicago: University of Chicago Press, 1967).

13. The terms used to describe young scientists in training can be opaque to the general public. Graduate students are preparing for the Ph.D. degree, whereas postdoctoral fellows have obtained that degree and are pursuing a semi-independent apprenticeship as final preparation for appointment to the faculty of a college or university, or for employment by a research institute or commercial firm. Graduate studies in biomedical science typically require on the order of six years, postdoctoral studies a further three years or more (with the duration of postdoctoral fellowships increasing gradually, as research disciplines grow more sophisticated and the job market becomes more congested).

14. See also J. Michael Bishop, "The Discovery of Proto-oncogenes," *FASEB Journal* 10 (1996): 362–364.

15. John Henry Cardinal Newman, *The Idea of a University* (New Haven: Yale University Press, 1996), pp. 5–6.

16. See the autobiographical sketch by Harold E. Varmus in Frangsmyr, *Les Prix Nobel*, pp. 215–219.

17. Jonathan Weiner, *Time, Love, Memory* (New York: Vintage, 1999), p. 68.

18. Blaise Pascal, *Lettres Provinciales*, as quoted in ibid., p. 71.

19. As quoted in Alan Packer, "The Permanent Postdoc," *Nature Genetics* 30 (2002): 11.

20. Typically iconoclastic, George Bernard Shaw saw things otherwise: "In art

the highest success is to be the last of your race, not the first. Anybody, almost, can make a beginning: the difficulty is to make an end—to do what cannot be bettered." As quoted by James M. Keller, in notes on the "Four Last Songs" of Richard Strauss," *Playbill* 4., no. 6 (February 2002): 27B. But why not make a beginning that cannot be bettered in the long run?

21. Ben Shahn, *The Shape of Content* (Cambridge: Harvard University Press, 1985), p. 19.

22. Crick, *What Mad Pursuit*, p. 145.

23. A prevalent example is the difficulty that women have encountered in trying to establish careers in science. The U.S. record in this regard is lamentable and improving only gradually. For narrative accounts, see Elga Wasserman, *The Door in the Dream: Conversations with Eminent Women in Science* (Washington, D.C.: Joseph Henry Press, 2000).

24. Herbert Muschamp, "Interior City: Hotel as the New Cosmopolis," *New York Times*, October 5, 2000, p. B8.

25. Freeman Dyson, *From Eros to Gaia* (New York: Pantheon, 1992), p. 191.

26. Ibid., p. 197.

27. G. H. Hardy, *A Mathematician's Apology* (Cambridge, Eng.: Cambridge University Press, 1981), p. 77.

28. A ringing defense of ambition can be found in Joseph Epstein, *Ambition* (Chicago: Ivan R Dees, 1989), and a searching analysis of its psychological genesis in Ernest Becker, *The Denial of Death* (New York: Free Press, 1973).

29. The quote from Fats Waller is anecdotal. I have never been able to authenticate it, but it seemed too apt to omit.

30. As quoted in Lewis Wolpert, *The Unnatural Nature of Science* (Cambridge: Harvard University Press, 1993), p. 57.

31. The account of Ramon Y Cahal, and his quote, are from William R. Everdell, *The First Moderns* (Chicago: University of Chicago Press, 1997), pp. 110–111.

32. The Joint Steering Committee was originally convened and for some years chaired by Professor Marc Kirschner, once my colleague at UCSF but now in exile at Harvard Medical School. His successors as chair have been Eric Lander, from the Massachusetts Institute of Technology, and more recently, Harold Varmus. Tom Pollard, now at Yale University, organized and still directs the team of more than two thousand correspondents. Elizabeth Marincola has served throughout as executive director. My own role has been modest, serving as a member of the committee and as program advisor to the Congressional Biomedical Research Caucus.

33. Several of us eventually took it upon ourselves to tell the newly elected administration of President Bill Clinton what it needed to do on behalf of biomedi-

cal research—see J. Michael Bishop, Marc Kirschner, and Harold Varmus, "Science and the New Administration," *Science* 259 (1993): 444–445. Had Harold Varmus not soon become director of the NIH, I presume that all of our recommendations would have fallen on deaf ears. As it was, Harold had to accommodate himself to realities that our manifesto ignored.

34. Ambrose Bierce, *The Devil's Dictionary* (New York: Dover, 1993), p. 95.

35. The general reader can be forgiven for knowing no more than that anonymous congressman about the distinction between the National Institutes of Health (NIH) and the National Science Foundation (NSF). Simply put, the NIH is the principal federal agency responsible for both fundamental and applied research that relates to human health. In contrast, the NSF distributes federal funds for research in a broad range of disciplines, including biology, mathematics, social and behavioral sciences, physics, and engineering. Such research has deep implications for many aspects of our society, including health, defense, and commerce. Chapter 5 discusses the missions of the two agencies in greater detail and addresses the unfortunate imbalance of federal appropriations between the two (with NIH faring much better than NSF).

36. Alan Ehrenhalt, "Another Chance to Make the Sale," *New York Times,* June 6, 2001, p. A31.

37. Bill Keller, "Up with Moguls! Exploit the Rich!" *New York Times,* November 3, 2001, p. A23.

38. Chester Bowles made these remarks during an informal seminar with college students. The occasion was reported to me by Bruce Alberts, president of the U.S. National Academy of Sciences.

39. Adrienne Rich, *The Dream of a Common Language* (New York: W. W. Norton, 1978), p. 67.

40. Henry Rosovsky, *The University: An Owner's Manual* (New York: W. W. Norton, 1990), p. 20.

41. For the sake of full disclosure, I must report that I have retained some semblance of the professoriate: I still deliver a series of lectures to medical students (see Chapter 3 for an expanded example), and I have maintained a research program, modest in size but immodest in ambition.

42. Seymour Benzer, as quoted in Weiner, *Time, Love, Memory,* p. 45.

43. As quoted by Denis Donoghue in "The Myth of W.B. Yeats," *New York Review of Books,* February 19, 1998, pp. 17–19. The quotation is on p. 19.

44. Kerr, *Gold and the Blue,* p. 326.

45. The biotechnology industry is based on the technique known as recombinant DNA, the invention of which is largely credited to professors Herbert Boyer

of UCSF and Stanley Cohen of Stanford University—they hold the patents on this technology. The industry got its proper start with the founding of Genentech, Inc., by Boyer and a venture capitalist, the late Robert Swanson, in April of 1976.

46. There are exceptions to this reluctance. California governor Gray Davis has made the funding of new research institutes at the University of California a centerpiece of his legislative agenda.

47. Malcolm Gladwell, "Brain Trust," *New Yorker,* February 23 and March 2, 1998, p. 121.

48. Anne Matthews, *Bright College Years* (Chicago: University of Chicago Press, 1998), pp. 19 and 150.

49. As quoted in Ragnar Sohlman and Henrik Schuck, *Nobel: Dynamite and Peace* (New York: Cosmopolitan Book Corporation, 1929), p. 249.

3. People and Pestilence

1. For further reading, consult Jared Diamond, *Guns, Germs, and Steel* (New York: W. W. Norton, 1997); William H. McNeill, *Plagues and People* (Garden City, N.Y.: Anchor, 1976); and MacFarlane Burnet and David O. White, *Natural History of Infectious Disease* (Cambridge, Eng.: Cambridge University Press, 1972).

2. The description of plague is based in part on Barbara W. Tuchman, *A Distant Mirror* (New York: Ballantine, 1978), pp. 92–125. The story is told in greater detail by Philip Ziegler in his esteemed book *The Black Death* (New York: Penguin, 1982). For a more recent account of the plague and, in particular, its consequences, see Norman F. Cantor, *In the Wake of the Plague* (New York: Free Press, 2001).

3. Until recently, the microbial causes of ancient epidemics have been inferred with little confidence from historical records of clinical manifestations, geographical distribution, and other such clues. But recent technological advances have made it possible to examine archaeological specimens for the residues of microbes and, thus, to make credible attributions of cause. For an account of how this strategy is being applied to study the Plague of Justinian, see Daniel Del Castillo, "A Long Ignored Plague Gets Its Due," *Chronicle of Higher Education,* February 15, 2002, pp. A22–A23.

4. Tuchman, *Distant Mirror,* pp. 92–93.

5. Giovanni Boccaccio, *The Decameron* (New York: Mentor, 1982), p. 9.

6. Bernal Diaz del Castillo, *The Discovery and Conquest of Mexico* (New York: Farrar, Straus, and Giroux, 1956), describes the conquest of the Aztecs, as recorded by one of the Spanish conquistadors.

7. Contemporary scholars are presently disputing the exact size of the Meso-american population at the time of the Spanish landing, but no one has challenged the conclusion that the population declined drastically as a result of infectious diseases introduced by the European invasion.

8. For a history of syphilis, see Theodor Rosebury, *Microbes and Morals* (New York: Viking, 1971), pp. 23–93.

9. I have been using this quotation for years but have long since misplaced its source. I am not alone in trusting its authenticity, however: it has appeared recently in other venues (but also without attribution of provenance)—see, for example, Jonathan B. Tucker, *Scourge: The Once and Future Threat of Smallpox* (New York: Atlantic Monthly Press, 2001), p. 11.

10. Boccaccio, *Decameron*, pp. 6–7.

11. An extensive description of Girolamo Fracastoro and his ideas can be found in Rosebury, *Microbes and Morals*, esp. pp. 25–48.

12. Ibid., p. 40. The term "germ" was an abstraction for Fracastoro. He had no direct knowledge of the microbial world.

13. John Hunter, in a letter to his student Edward Jenner, as quoted in John Kobler, *The Reluctant Surgeon: A Biography of John Hunter* (Garden City, N.Y.: Doubleday, 1960), p. 175.

14. Lawrence K. Altman, *Who Goes First?* (New York: Random House, 1986), p. 7.

15. Ibid., pp. 7–8; George Qvist, *John Hunter, 1728 1793* (London: Heinemann Medical Books, 1981), pp. 45–53.

16. Elie Metchnikoff, as quoted in Richard M. Krause, "Metchnikoff and Syphilis Research during a Decade of Discovery, 1900–1910," *ASM News* 62 (1996): 307–310. The quotation is on p. 309.

17. For an accessible account of Semmelweis's findings and struggles, see Hall Hellman, *Great Feuds in Medicine* (New York: John Wiley and Sons, 2001), pp. 33–52.

18. Translated from Ignac P. Semmelweis, *Zwei Offene Briefe An Hofrath Dr. E. C. J. Von Siebold Und An Hofrath Dr. F. W. Scanzoni* (Pest, 1861), p. 40.

19. As quoted in Frank G. Slaughter, *Immortal Magyar* (New York: Henry Schuman, 1950), p. 177.

20. Ignac P. Semmelweis, *The Etiology, the Concept, and the Prophylaxis of Childbed Fever,* trans. K. Codell Carter (Madison: University of Wisconsin Press, 1983).

21. Sherwin B. Nuland, "The Enigma of Semmelweis—An Interpretation," *Journal of the History of Medicine* 34 (1979): 255–272, and Codell Carter, Scott Abbott, and James L. Siebach, "Five Documents Relating to the Final Illness and

Death of Ignaz Semmelweis," *Bulletin of the History of Medicine* 69 (1995): 225–270. An alternative and longstanding view is that Semmelweis died of an infection with streptococcus, his microbial adversary on the obstetrical wards—see Slaughter, *Immortal Magyar.*

22. Slaughter, *Immortal Magyar,* p. 206.

23. Gerald Weissman, *Democracy and DNA* (New York: Hill and Wang, 1995), p. 26.

24. For a biography of Joseph Lister, see Richard B. Fisher, *Joseph Lister* (New York: Stein and Day, 1977).

25. For a contemporary biography of Louis Pasteur that recounts his alleged misconduct as a scientist, see Gerald Geison, *The Private Science of Louis Pasteur* (Princeton, N.J.: Princeton University Press, 1995). For a more admiring and romanticized view of the great man, see Renee J. Dubos, *Louis Pasteur* (Boston: Little, Brown, 1950). And for a vigorous defense of Pasteur against the accusations by Geison, see Max Perutz, "The Pioneer Defended," *New York Review of Books,* December 21, 1995, pp. 54–58, reprinted as "Deconstructing Pasteur" in *I Wish I'd Made You Angrier Earlier* (Plainview, N.Y.: Cold Spring Harbor Laboratory Press, 1998), pp. 119–130.

26. Geison, *Private Science of Louis Pasteur,* p. 139.

27. Ibid.

28. Thomas D. Brock, *Robert Koch: A Life in Medicine and Bacteriology* (Madison, Wis.: Science Tech, 1988), p. 4.

29. Emmy Koch, as quoted in ibid., p. 31.

30. Ibid., p. 233.

31. Another version of this story, not well documented and perhaps apocryphal, credits Koch's wife Emmy with the suggestion of agar, whose coagulative properties she knew from her experience with preparing Japanese cuisine. The likelihood that Emmy had encountered Japanese cuisine in Wollstein strikes me as small. There is no trace of it there now, by any account. But I am favorably disposed to any effort that might make Emmy's contribution to medical progress more visible.

32. Details of this argument can be found in Cantor, *In the Wake of the Plague.*

33. Mark Derr, "New Theories Link Black Death to Ebola-like Virus," *New York Times,* October 2, 2001, p. D4.

34. For a detailed description of Pasteur's trial of immunization against anthrax, see Geison, *Private Science of Louis Pasteur,* pp. 145–176. Geison's close reading of the archives suggests that Pasteur's triumph against anthrax was blighted by deception. Pasteur used one form of vaccine, but claimed to have used

another in order to strengthen his case for priority in the development of anthrax vaccine. In Geison's words, "The conclusion is unavoidable: Pasteur deliberately deceived the public, including especially those scientists most familiar with his published work, about the nature of the vaccine actually used [in the immunizations against anthrax]" (p. 156). There is no question, however, that the vaccine worked, or that Pasteur subsequently succeeded with the vaccine that he only purported to have used in the first case.

35. Brock, *Robert Koch*, p. 273.

36. Ibid., p. 171.

37. Stephen Jay Gould, *Full House* (New York: Three Rivers Press, 1996), p. 4.

38. For more about our normal flora, see Theodor Rosebury, *Life on Man* (New York: Viking, 1969).

39. W. H. Auden, from "A New Year Greeting," in *Epistle to a Godson and Other Poems* (New York: Random House, 1972), pp. 12–13.

40. The deliberate ingestion of cholera bacteria by Pettenkofer and his colleagues is considered to be among the earliest examples of informed self-experimentation. For a more extensive account and further references, see Altman, *Who Goes First?* pp. 23–28 and 324–325.

41. As quoted in Roy Porter, *The Greatest Benefit to Mankind* (New York: W. W. Norton, 1998), p. 437.

42. As quoted in Alfred S. Evans, "Two Errors in Enteric Epidemiology: The Stories of Austin Flint and Max von Pettenkofer," *Reviews of Infectious Diseases* 7 (1985): 434–440. The quotation is on p. 438. Also quoted in Altman, *Who Goes First?* p. 25.

43. The evidence for these origins is summarized in Diamond, *Guns, Germs, and Steel*, pp. 195–210.

44. For a discussion of how and why the virulence of microbes evolves, see Paul W. Ewald, "The Evolution of Virulence," *Scientific American* (April 1993): 86–93.

45. An account of myxomatosis virus and Murray Valley encephalitis can be found in the autobiography of MacFarlane Burnet, *Changing Patterns* (New York: American Elsevier, 1968), chap. 9, pp. 105–120.

46. The argument that plague set the stage for the Renaissance is laid out in detail by David Herhily, *The Black Death and the Transformation of the West* (Cambridge: Harvard University Press, 1997). The quotations are from p. 38.

47. For more about Florence Nightingale as scientist, see Paul D. Stolley and Tamar Lasky, *Investigating Disease Patterns* (New York: Scientific American Library, 1995).

48. Ibid., p. 42.

49. Claire M. Fagin and Donna Diers, "Contemporary Nightingales," *New York Times*, November 7, 2000, p. D7.

50. For a concise account of work of John Snow and illustrations of some of his data, see Stolley and Lasky, *Investigating Disease Patterns*, pp. 33–39.

51. For a revisionist view of John Snow and the Broad Street pump, see Donald Cameron and Ian G. Jones, "John Snow, the Broad Street Pump and Modern Epidemiology," *International Journal of Epidemiology* 12 (1983): 393–396. The quote is from Jessica Ludwig, "UCLA Epidemiologist Creates a Web Site about a Pioneer in the Field," *Chronicle of Higher Education*, chronicle.com/infotech, June 13, 2000.

52. John Snow, as quoted in Cameron and Jones, "John Snow," p. 393.

53. John Snow, as quoted in ibid., p. 394.

54. As quoted in Porter, *Greatest Benefit to Mankind*, p. 412.

55. For a consideration of smallpox as a biological weapon, see Tucker, *Scourge*.

56. As quoted in Richard Preston, "The Demon in the Freezer," *New Yorker*, July 12, 1999, p. 52.

57. Thomas Jefferson, in a letter to Edward Jenner, as quoted in Kobler, *Reluctant Surgeon*, p. 182. Jefferson was an energetic proponent of vaccination against smallpox. In 1801, he personally vaccinated two hundred of his neighbors in Virginia, and in 1803, he sent smallpox vaccine west with Meriwether Lewis and William Clark so that they might vaccinate Indians while exploring the Louisiana Purchase. The vaccination campaign was foiled by spoilage of the vaccine.

58. There has been at least one subsequent documented death from smallpox—a forty-year-old medical photographer at the medical school in Birmingham, England, who became infected with virus that had escaped from a research laboratory. The scientist responsible for the laboratory later committed suicide. The episode is described in detail by Tucker, *Scourge*, pp. 124–132.

59. George Bernard Shaw, as quoted in ibid., p. 34.

60. For a full critique of the safety and efficacy of Pasteur's original vaccine against rabies, see Geison, *Private Science of Louis Pasteur*.

61. See Perutz, "Pioneer Defended."

62. An account of Pasteur's apparent success with Joseph Meister can be found in Dubos, *Louis Pasteur*, pp. 335–336.

63. Marie-Louise Pasteur, as quoted in Robert L. Krasner, "Pasteur: High Priest of Microbiology," *ASM News* 611 (1995): 575–579. The quotation is on p. 578.

64. Geison, *Private Science of Louis Pasteur,* p. 192.

65. Egon Gartenberg, *Mahler* (New York: Schirmer, 1978), p. 178.

66. The life and work of Gerhard Domagk is described in Frank Ryan, *The Forgotten Plague* (Boston: Little, Brown, 1992), chaps. 6–8, pp. 75–130.

67. As quoted in ibid., p. 97.

68. As quoted in ibid., pp. 102–103.

69. For a full account of the discovery and development of penicillin, see Gwyn Macfarlane, *Alexander Fleming* (Cambridge: Harvard University Press, 1984). The exact manner in which the historic petri dish was produced is not known. Fleming himself published contradictory accounts (see pp. 117–126 of the Macfarlane book). A more succinct reconstruction of how Fleming's discovery may have come about is in Max F. Perutz, *Is Science Necessary?* (New York: Dutton, 1989), pp. 154–156.

70. The first use of the aphorism is described in Geison, *Private Science of Louis Pasteur,* p. 147. The aphorism was displayed prominently in the entryway to the residential hall for Harvard Medical School when I was a student there. For two years, I walked beneath it several times each day. At the time, it seemed an abstraction to me. But eventually, I was fortunate enough to see it made real by my own experience. (Chapter 4 tells that story.) I recently revisited the hall for the first time since my graduation. The inscription is still there, but what I had not remembered is that it is written in French, a language that I have never learned. Someone must have given me the translation early in my tenure as a medical student. The thought was a powerful presence in my life then, taunting me because I did not expect to have the sort of chance that Pasteur had in mind. Pasteur also used another aphorism, taken from Virgil: "Luck comes to the bold." And indeed, no one could accuse Pasteur of timidity.

71. Macfarlane, *Alexander Fleming,* p. 174.

72. Gwyn Macfarlane, *Howard Florey* (Oxford, Eng.: Oxford University Press, 1979), p. 331.

73. For an account of penicillin and patents, see Gwyn Macfarlane, *Alexander Fleming,* pp. 205–206.

74. It is now clear that streptomycin was actually discovered by Albert Schatz, a Ph.D. student with Waxman. Although he was listed as codiscoverer on the patent for streptomycin, Schatz was given no public credit for the discovery until his belated lawsuit over patent rights forced the issue and gained him a share of the royalties. In addition to royalties, the suit earned Schatz hostility among the scientific community. The Nobel committee passed him over. He was not honored until 1994, when Rutgers University, where he and Waxman had done their historic

work, awarded him the Rutgers Medal as "codiscoverer" of streptomycin. For details, see Burton Feldman, *The Nobel Prize* (New York: Arcade, 2002), pp. 276–279.

75. The potential effect of this advance on the study of infectious diseases is illustrated in miniature by the recent decoding of the entire genome of the cholera bacterium (see Victor J. DiRita, "Genomics Happens," *Science* 289 [2000]:1488–1489). Fully half of the bacterial genes uncovered by this analysis have no known function, many have not even been encountered before in any context, and some of them are likely to account for the ability of the bacterium to cause disease. John Snow and Robert Koch would have been intrigued; Max Pettenkofer might well still be scoffing. The genome of the plague bacillus has also been decoded, revealing a record of biological mobility and adaptation. The plague bacillus was once an innocuous inhabitant of the gastrointestinal tract. But its genome has been peppered by genes acquired by horizontal transfer from other creatures, freeing the microbe of its old habitat in the gastrointestinal tract, arming it with the ability to colonize insects, and conferring the ability to infect mammalian cells—in aggregate, the ingredients for biological catastrophe. A general account of these remarkable findings can be found in Stuart T. Cole and Carmen Buchrieser, "A Plague o' Both Your Hosts," *Nature* 413 (2001): 467–470.

4. Opening the Black Box of Cancer

The remarks in the epigraph by Senator Neely are taken from an address made on the floor of the U.S. Senate in 1928, in support of the first effort to pass a bill for cancer research. The bill would have appropriated $100,000. It passed the Senate but failed in the House of Representatives. Congress eventually established the National Cancer Institute in 1937. The annual budget for the institute is now more than $4 billion.

1. Peyton Rous, "The Challenge to Man of the Neoplastic Cell," in *Les Prix Nobel En 1966* (Stockholm: P. A. Norstedt and Sons, 1967), pp. 162–171. The quotation is from p. 162.

2. For an account of Robert Hooke, see John A. Moore, *Science as a Way of Knowing* (Cambridge: Harvard University Press, 1993), pp. 97–101 and 253–255. The book provides a graceful narrative of the entire history of cell research.

3. Ibid., p. 99. The title rightfully refers to the instruments as "magnifying glasses." The early "microscopes" used by Leeuwenhoek, Hooke, and many others employed only a single lens. The true microscope is "compound," utilizing two lenses in sequence.

4. Ibid., p. 261.

5. Edmund B. Wilson, *The Cell in Development and Inheritance* (New York: Macmillan, 1896), p. 4.

6. The roles of Virchow, Waldeyer, and others in generating a cellular understanding of cancer are recounted by L. J. Rather, *The Genesis of Cancer* (Baltimore: Johns Hopkins University Press, 1978).

7. The term "genome" was devised to describe the aggregate of all genes in an individual virus or organism. Each gene is encoded in a stretch of the stringlike molecule known as DNA. But genes occupy only a small portion of all the DNA in human cells. Some of the remaining DNA is involved in the regulation of genes, some is parasitic, some of it may be adventitious, and some may have presently unknown purpose(s). The genome of bacteria/prokaryotes is typically (but not always) contained on a single molecule of DNA, whereas the genome of most eukaryotes and of humans in particular is distributed among multiple chromosomes, each of which contains a separate, large molecule of DNA. The recently devised term "genomics" refers to the charting and decoding of genetic information, as exemplified by the Human Genome Project, and the use of that information to study biological processes, both normal and abnormal.

8. Frank G. Slaughter, *Immortal Magyar: Semmelweiss, Conquerer of Childbed Fever* (New York: Henry Schuman, 1950), p. 174.

9. For an accessible account of how cancer arises and progresses, see Harold Varmus and Robert A. Weinberg, *Genes and the Biology of Cancer* (New York: Scientific American Library, 1993).

10. Theodosius Dobzhansky, "Nothing in Biology Makes Sense Except in the Light of Evolution," *American Biology Teacher* 35 (1973): 125–129.

11. Richard Doll and Richard Peto, *The Causes of Cancer* (Oxford, Eng.: Oxford University Press, 1981).

12. The term "carcinogenesis" originally referred only to the production of "carcinomas," one of several forms of malignant tumors. But it is now loosely used to describe all forms of tumorigenesis, generally with an implication of external causes. Those external causes are in turn referred to as "carcinogens."

13. More about Wilhelm Hueper can be found in Robert N. Proctor, *Cancer Wars* (New York: Basic Books, 1995), pp. 36–48.

14. I am indebted to Dr. James Miller for providing me the illustration and translation of the haiku by Yamagiwa.

15. Elof Axel Carlson, *Genes, Radiation, and Society: The Life and Work of H. J. Muller* (Ithaca, N.Y.: Cornell University Press, 1981), p. 174.

16. Walter S. Sutton, "The Chromosomes in Heredity," *Biological Bulletin* 4 (1903): 24–39.

17. The account of Walter Sutton is based on Moore, *Science,* pp. 304–314.

18. Theodor Boveri, *The Origin of Malignant Tumors* (Baltimore: Williams and Wilkins, 1929), translation by Marcella Boveri. The quotations are from pp. 26–27.

19. For a personal account of the discovery of tumor viruses, see Rous, "Challenge to Man."

20. Leon Edel, *Henry James: A Life* (New York: Harper and Row, 1985), p. 671.

21. It was Francis Crick who coined the term "central dogma." After the discovery of reverse transcriptase, he pointed out that the original formulation of this dogma had not precluded the transfer of information from RNA to DNA, only that from protein to RNA or DNA. The latter proscription has held to this day and is not likely to fall. For more about the discovery of reverse transcriptase, see Chapter 2.

22. Rous, "Challenge to Man," p. 167.

23. As quoted in George Klein, *The Atheist and the Holy City* (Cambridge: MIT Press, 1990), p. 122.

24. As quoted in Walter Gratzer, *A Literary Companion to Science* (New York: W. W. Norton, 1990), p. 59.

25. Rous's self-assurance proved useful to Ernst Wynder. When Wynder came under fire from the director of the Sloan-Kettering Institute for offending the tobacco industry, Rous weighed in on Wynder's behalf and carried the day.

26. The dispensability of *SRC* from the virus had been shown most clearly by our collaborator Peter Vogt, who was then working at the University of Southern California. Virus from which *SRC* had been eliminated was a crucial reagent in our experiments. Peter provided this virus to us, along with many other assists, and joined us as an author of the paper that announced the presence of *SRC* in normal cells. The extraordinarily fruitful collaboration with Peter exemplifies the collegial nature of science described in Chapter 2.

27. Dominique Stehelin, Open Letter to the Nobel Committee on Physiology or Medicine (November 10, 1989). In possession of the author.

28. Dominique Stehelin, Harold E. Varmus, J. Michael Bishop, and Peter K. Vogt, "DNA Related to the Transforming Gene(s) of Rous Sarcoma Virus Is Present in Normal Avian DNA," *Nature* 260 (1976): 70–73. The quotation is from p. 173.

29. There is no reason to believe that the pirating of cellular genes by retroviruses is limited to proto-oncogenes. But proto-oncogenes come to our attention because their presence in virus gives rise to tumors.

30. Rous, "Challenge to Man," p. 166.

31. Peyton Rous, "Surmise and Fact on the Nature of Cancer," *Nature* 183 (1959): 1357–1361.

32. Paul Broca is known to the general public through his starring role in the best-selling book by Carl Sagan, *Broca's Brain* (New York: Random House, 1974). The book makes no mention of Broca's extracurricular interest in familial cancer.

33. Some strict taxonomists distinguish genes for DNA repair from tumor suppressor genes. I have not chosen that course here. By suppressing mutations, DNA repair genes suppress the frequency of cancer—they do so indirectly, as explained in the text, but that does not make their role in tumor suppression any less vital.

34. In genetic parlance, mutations in proto-oncogenes are dominant: their pathogenic effects are felt even in the presence of a normal copy of the gene—evil overrides good. In contrast, mutations in tumor suppressor genes are usually recessive: their effects are felt only in the absence of a normal copy of the gene. Since our cells contain two copies of virtually all of our genes, it requires two separate genetic events to create a complete deficiency of a particular tumor suppressor gene. The general reader does not need to master these complexities for present purposes. A more complete consideration of genetic dominance and recessiveness is given in Chapter 5.

35. Health and medical science were a recurrent theme of Rivera's mural painting. For an account of the Diego Rivera mural in Rockefeller Center, see Patrick Marnham, *Dreaming with His Eyes Open* (New York: Alfred A. Knopf, 1998), pp. 248–260. Rivera made a few modifications to the original mural when reconstructing it in Mexico City. The most telling was to place a likeness of John D. Rockefeller, Jr., just beneath a swarm of syphilis bacteria—a symbolic retaliation for the cruel treatment of Rivera's work in Rockefeller Center.

36. Norman Mailer, *Tough Guys Don't Dance* (New York: Ballantine Books, 1985), p. 257.

37. Cells committing altruistic suicide contain a structural image evocative of falling petals. The word "apoptosis" is constructed from Greek to denote that image.

38. Chapter 5 contains a more extensive discussion of the ethical and practical conundrums posed by genetic testing.

39. The retrospective detection of Hubert Humphrey's bladder cancer is described by Ralph H. Rubren, Peter Van Der Riet, Yener S. Erozan, and David Sidransky, "Brief Report: Molecular Biology and the Early Detection of Carcinoma of the Bladder—The Case of Hubert H. Humphrey," *New England Journal of Medicine* 330 (1994): 1276–1278.

40. H. G. Wells, as quoted in Walter Gratzer, "Gardner's Choice," *Nature* 313 (1984): 605. The quote comes originally from an obscure novel by Wells, *Meanwhile*.

41. Susan Sontag, *Illness as Metaphor* (New York: Farrar, Straus and Giroux, 1978), pp. 20, 68, and 87.

42. Robert Frost, "Kitty Hawk," in *In the Clearing* (New York: Holt, Rinehart and Winston, 1962), p. 56.

5. Paradoxical Strife

1. Edward O. Wilson has written of our innate respect for all of life in *Biophilia* (Cambridge: Harvard University Press, 1984).

2. Portions of Chapter 5 are adapted from lectures delivered to a forum convened by Sigma Xi and a Stated Meeting of the American Academy of Arts and Sciences. That delivered to the Sigma Xi forum was published as "Paradoxical Strife: Science and Society in 1993," in *Ethics, Values, and the Promise of Science* (Research Triangle Park, N.C.: Sigma Xi, 1993), pp. 95–114; the one given to the academy was published as "Paradoxical Strife: Science and Society," in *Bulletin of the American Academy of Arts and Sciences* 48 (1995): 10–30. Excerpts also appeared as "The Crisis of Contemporary Science: Enemies of Promise," *Wilson Quarterly* 19 (Summer 1995): 61–65. The "strife" of which I write here arises when the capabilities of science and their technological offspring challenge human values and beliefs. I have not addressed a broader range of adverse effects that can arise from science and technology, such as acid rain, global warming, and extinction of species. For a consideration of such effects, see Edward Tenner, *Why Things Bite Back* (New York: Knopf, 1996).

3. The vigor of the opposition to UCSF obscured the fact that most of the neighborhood was oblivious to the debate. Opinion polls indicated that only a small minority of local residents was either aware of the issues or seemed to care very much about them. I make this point not to trivialize the opposition, but to emphasize how frequently public institutions must contend with minority views whose strength arises in part from majority complacency.

4. There are roughly three yards of DNA in each human cell and 300 trillion cells in each human body—see Chapter 4.

5. As quoted in Freeman J. Dyson, "Science in Trouble," *American Scholar* 62 (1993): 513–525. The quotation is on p. 517.

6. Bernard D. Davis, *Storm over Biology* (Buffalo, N.Y.: Prometheus Books, 1986), p. 243.

7. The confrontation between UCSF and its neighbors is described in detail by Charles Piller, *The Fail-Safe Society* (Berkeley: University of California Press, 1991), chap. 5, pp. 118–157.

8. Ibid., p. 152.

9. Ibid., p. 135.

10. The prevalence of single-gene defects in the population is made possible by our genetic constitutions. We are "diploid" organisms: with the exception of genes carried on the sex chromosomes in males, all of our genes are represented by two copies in our cells—one derived from our mothers, the other from our fathers. Virtually all single-gene deficiencies are "recessive": they have a deleterious effect only if both copies of the gene are defective. Individuals carrying only one defective copy of a gene are known as "heterozygotes" and are usually asymptomatic "carriers" of the genetic predisposition to disease; those with two copies are "homozygotes" and are predisposed to develop the disease. Homozygotes arise only if both parents carry at least one defective copy of the pertinent gene. The rarity with which two heterozygotes for the same mutation find one another and mate accounts in large part for the relative infrequency of disease due to single-gene defects in the general population.

11. Chapter 3 describes the manner in which microbes and their hosts often evolve toward mutual compatibility, helping to account for the relative infrequency of disease in response to infection by many agents.

12. Chapter 3 describes an example of how genes influence susceptibility to infection with HIV.

13. Alex Mauron, "Is the Genome the Secular Equivalent of the Soul?" *Science* 291 (2001): 831–832. The quotation appears on p. 832.

14. Jonathan Weiner, *Time, Love, Memory* (New York: Random House, 1999), p. 66.

15. For a survey of the prospects and difficulties of genetic testing, see Neil A. Holtzman, *Proceed with Caution* (Baltimore: Johns Hopkins University Press, 1989).

16. The screening program in Sweden and its effects were reviewed by Eric A. Wulfsberg et al., "Alpha-1-Antitrypsin Deficiency: Impact of Genetic Discovery on Medicine and Society," *Journal of the American Medical Association* 271 (1994): 217–222.

17. For an accessible review of legal and legislative issues arising from genetic testing, see Philip R. Reilly, "Legal Issues in Genomic Medicine," *Nature Medicine* 7 (2001): 268–271.

18. See Michelle Andrews, "Genetic Tests Abound: Why Won't Insurers Pay?" *New York Times,* May 19, 2002, p. B9.

19. The action against the Burlington Northern Santa Fe Railway Company was reported by Tamar Lewin, "Commission Sues Railroad to End Genetic Testing in Work Injury Cases," *New York Times,* February 10, 2001, p. A7.

20. Marilyn Lewis Hampton, "Mother to Be's Painful Choice," *San Francisco Examiner*, November 8, 1992, p. E1.

21. Genetic testing is only one of several developments that have revived the discussion of eugenics. Far more contentious is the prospect that some day soon we may be able to directly alter the human genome, eliminating deleterious traits and installing desirable ones. For an authoritative account of eugenics and the issues it poses, see Daniel J. Kevles, *In the Name of Eugenics: Genetics and the Uses of Human Heredity* (New York: Knopf, 1985). Ironically, the eugenics movement was founded by a cousin of Charles Darwin, Francis Galton.

22. Max Perutz, "Should Genes Be Screened?" *New York Review of Books*, May 18, 1989, pp. 34–36. The quotation is on p. 36.

23. In vitro fertilization could also be used to create test-tube embryos specifically for the purpose of harvesting stem cells, but this approach has not figured much in the current disputes over stem cell research and would surely offend the opponents of such research. As explained in the text, however, embryos created by "therapeutic cloning" are another matter entirely.

24. Richard Doerflinger, quoted in the *New York Times*, August 15, 2001, p. A18.

25. For a discussion of this issue, see Bert Vogelstein, Bruce Alberts, and Kenneth Shine, "Please Don't Call It Cloning!" *Science* 295 (2002): 1237. The authors propose the substitution of "nuclear transplantation" for "therapeutic cloning." The debate over therapeutic cloning has produced a number of alternative terms, each crafted to respond to political or ethical sensitivities.

26. Opposition has also arisen from two other quarters: activists who fear that therapeutic cloning would represent the first step down a "slippery slope" leading eventually to eugenics; and feminists who fear "commodification" of human eggs. Advocates of the cloning answer that both of these potential problems can be regulated effectively by appropriate legislation.

27. Paul Nurse, as quoted in Geoff Dyer, "Norman Mailer Lends Weight to U.S. Anti-Cloning Coalition," *Financial Times*, April 1, 2002, p. 12.

28. I first encountered this suggestion in Hal Hellman, *Great Feuds in Science* (New York: John Wiley and Sons, 1998), pp. 160–161.

29. My upbringing as the son of a minister had little influence on my receptivity to evolution. My father was both liberal and unassuming about biological facts. Then again, I cannot recall hearing anything about evolution until I entered college. There I studied comparative anatomy and saw the reality of evolution made plain. It was an exhilarating and important step in my intellectual development, and my first experience of aesthetic pleasure from the coherence of science.

30. See "The Pope's Message on Evolution and Four Commentaries," *Quarterly Review of Biology* 72 (1997): 382–383.

31. For details of the Scopes Trial, see Edward J. Larson, *Summer for the Gods* (Cambridge: Harvard University Press, 1997). To this day, popular knowledge of the trial is based mainly on the play *Inherit the Wind*, by Jerome Lawrence and Robert E. Lee. But the playwrights readily conceded that the play was "not journalism," but rather a fictionalized account designed to address the dangers posed by Senator Joseph R. McCarthy and his infamous hearings, a chapter later in American history than the Scopes Trial.

32. Larson, *Summer for the Gods*, p. 89.

33. The universe is thought to be at least 10 billion years of age, and the earliest forms of life to be found so far apparently existed at least 3.5 billion years ago. There is no credible estimate for when life actually originated, but it was unlikely to have been more than 4.5 billion years ago, since that is the estimated age of the planet Earth. I ignore here the possibility that life might have arrived on Earth in some primitive form from elsewhere in the universe, a scenario dubbed "panspermia" in 1906 by the Swedish physicist Svante Arrhenius, advocated by the great cosmologist Sir Fred Hoyle, and even given credence by Francis Crick in a variant that he called "directed panspermia," in which the primitive form of life arrived in a spaceship launched by an advanced civilization elsewhere in the universe—see Francis Crick, *Life Itself* (New York: Simon and Schuster, 1981).

34. For a critical overview of "intelligent design theory" and its protagonists, see Frederick Crews, "Saving Us from Darwin, Parts I and II," *New York Review of Books*, October 4, 2001, pp. 24–27, and October 18, 2001, pp. 51–55.

35. The quarrel between postmodernism and the natural sciences is recounted in Paul R. Gross and Norman Lefitt, *Higher Superstition* (Baltimore: Johns Hopkins University Press, 1994). It might also help if those inclined to take postmodernism seriously read Frederick Crews, *Postmodern Pooh* (New York: North Point Press, 2001).

36. For a description of the hoax, its context, and its aftermath, see Alan Sokal and Jean Bricmont, *Fashionable Nonsense* (New York: Picador, 1998).

37. This seems a suitable point to acknowledge that I have used the words academe and academy interchangeably. But Ambrose Bierce saw the matter otherwise, defining academe as "an ancient school that taught morality and philosophy" and the academy as "a modern school that teaches football." It seems to me that the nonsense known as postmodernism could only have arisen from Bierce's academy. See Ambrose Bierce, *The Devil's Dictionary* (New York: Dover, 1993), p. 3.

38. This and subsequent quotes from Vaclav Havel are recorded in Gerald Holton, "The Value of Science at the 'End of the Modern Era,'" in *Ethics, Values and the Promise of Science* (Research Triangle Park, N.C.: Sigma Xi, 1993), pp. 127–128. Havel's great cachet has given him access to prominent venues for his views, including the *New York Times*, the *New York Review of Books*, and numerous addresses to general and academic audiences—including a commencement address at Harvard University in which Havel laid out his disaffection with science in no uncertain terms.

39. These quotations are from Edward Rothstein, "Myths about Genius," *New York Times*, January 5, 2002, p. A17.

40. George E. Brown, "Rational Science, Irrational Reality," *Science* 256 (1992): 200–201.

41. Report of the Task Force on the Health of Research, 1992: Chairman's Report to the Committee on Science, Space and Technology, U.S. House of Representatives, 102d Cong., 2d sess. (Washington, D.C.: U.S. Government Printing Office, 1992), p. 14.

42. The term "peer review" is used as shorthand for the process by which scientists judge each other's competitive applications for research funds from the federal government and other sources. Pains are taken to instill rigor in the process, to avoid both favoritism and undue critical bias, and to keep the process as open to younger scientists as it is to their seniors. Peer review in the United States has its critics—no such human practice could be perfect. But the admiration that it commands throughout the international research community is perhaps the best testimony to its efficacy.

43. Peter Medawar, *Pluto's Republic* (Oxford, Eng.: Oxford University Press, 1985), p. 2.

44. Bernard Davis, unpublished manuscript in the author's possession, personal communication.

45. Brown, "Rational Science," p. 201.

46. Estimates of return on biomedical research run as high as twentyfold. But the reliability of these conclusions remains in dispute. See David Malakoff, "Does Science Drive the Productivity Train?" *Science* 289 (2000): 1274–1276.

47. As quoted in Washington Fax, June 4, 2001.

48. Brown, "Rational Science," p. 200. I was a strong admirer of Congressman Brown throughout his political career, so I easily forgave the annoyance of his task force.

49. Stephen W. Hawking, *A Brief History of Time* (New York: Bantam, 1988), p. 13.

50. Jared Diamond, *The Third Chimpanzee* (New York: Harper Perennial, 1993), p. 336.

51. U.S. Commission on National Security for the Twenty-first Century, "Road Map for National Security: Imperative for Change," as described in *Nature* 409 (2001): p. 651. The commission's concern extended to private enterprise as well, where investment in research was perceived to be losing ground even more severely.

52. Freeman Dyson, *From Eros to Gaia* (New York: Pantheon, 1992), p. 9.

53. Ibid., p. 26.

54. Steven Weinberg, *Dreams of a Final Theory* (New York: Pantheon, 1992), p. 274.

55. Ibid., p. 273.

56. Health centers associated with universities were once an exception, using portions of their surplus revenues to subsidize research and education. But most of these centers are now in perilous financial straits, and accordingly, their monetary contributions to research are dwindling.

57. This text was virtually complete before the September 11, 2001, terrorist attacks on the United States and the overt threat of bioterrorism that emerged soon thereafter. These events have necessitated the redirection of large federal resources to defense and recovery. But they also dramatize the need for improvements of technology in many venues, such as intelligence, security, construction, aircraft, microbiology, and public health—improvements that will only come from further research and development.

58. Mary Midgley, as quoted by Philip Clayton, "In Search of Unity," *Science* 409 (2001): 979.

59. Hawkings, *Brief History of Time*, p. 175.

60. Francis Crick, *The Astonishing Hypothesis: The Scientific Search for the Soul* (New York: Scribner's, 1994).

61. Dorothy Nelkin and M. Susan Lindee, *The DNA Mystique: The Gene as a Cultural Icon* (New York: Freeman, 1995).

62. James D. Watson, as quoted in Leon Jaroff, "The Gene Hunt," *Time*, March 20, 1989, pp. 62–67.

63. Walter Gilbert, as quoted in Richard Lewontin, *The Triple Helix* (Cambridge: Harvard University Press, 2001), p. 11.

64. The makeup of genes differs very little between chimpanzee and human. Yet the human brain is more complex and accomplished—the product of relatively rapid evolution. Recent work indicates that these distinctions may arise from differences in the regulation of genes: the pattern of gene expression—the

deployment of genetic function—differs substantially between the chimpanzee and human brain. For an account of this work, see Wolfgang Enard et al., "Intra- and Interspecific Variation in Primate Gene Expression Patterns," *Science* 296 (2002): 340–343.

65. Richard Feynman, *The Character of Physical Law* (Cambridge: MIT Press, 1993), pp. 125–126.

66. Larry Kramer, "Name an AIDS High Command," *New York Times*, November 15, 1992, p. A19. Kramer actually got his wish: the NIH did appoint a "high command" or "czar" to coordinate its research on AIDS. But more to my point, it was not long before scientists began to report the first drugs that could both improve and prolong the lives of individuals with AIDS.

67. As research on AIDS proceeded, most activists came to recognize the necessity of controlled clinical trials and turned their energies to procuring more funds for research.

68. For a general account of "Lorenzo's Oil," see Gina Kolata, "Lorenzo's Oil: A Movie Outruns Science," *New York Times*, February 9, 1993, p. B5–8. For a more sanguine view of the film than that offered here, see the editorial "Lorenzo Goes to Hollywood," *Nature Genetics* 3 (1993): 95–96.

69. For a brief account of the gene implicated in ALD and of the disappointing clinical trial of the oil therapy, see William R. Rizzo, "Lorenzo's Oil—Hope and Disappointment," *New England Journal of Medicine* 329 (1993): 801–802.

70. Arthur Caplan, as reported by Kolata, "Lorenzo's Oil," p. B8.

71. Daniel J. Kevles, *The Baltimore Case* (New York: W. W. Norton, 1998), p. 197. The book provides a comprehensive account of the congressional investigation of David Baltimore and Teresa Imanishi-Kari, and concludes by exonerating them both.

72. Alan Bloom, *The Closing of the American Mind* (New York: Simon and Schuster, 1987), p. 270.

73. Friedrich Nietzsche, as quoted in Hywel Williams, "University Challenge," *New York Times Literary Supplement*, January 22, 1993, p. 13.

74. A barely fictionalized Alan Bloom is the central character of the novel *Ravelstein* (New York: Viking, 2000) by none other than Saul Bellow.

75. As quoted in Bill Arnold and Rob Knies, "Beyond the Box Score," *San Francisco Chronicle*, May 13, 1989, p. E3.

76. Jill Kerr Conway, *A Woman's Education* (New York: Knopf, 2001), p. 140.

77. John Ciardi, "Fragment," in *The Collected Poems of John Ciardi* (Fayetteville: University of Arkansas Press, 1997), p. 365.

78. William Wordsworth, *Preface to the Lyrical Ballads* (Westport, Conn.: Greenwood Press, 1979), p. 124.

79. William Wordsworth, from "The Tables Turned," in *Wordsworth and Coleridge: Lyrical Ballads* (Oxford, Eng.: Oxford University Press, 1969), p. 105.

80. Lionel Trilling, *Mind in the Modern World* (New York: Viking, 1972), pp. 14, 41.

81. By way of example: the latest testing of science literacy in the United States scored as "proficient" for their grade level 29 percent of fourth graders, 32 percent of eighth graders, and only 18 percent of high school students. Paradoxically, the state that is home to the most Nobel laureates, California, scored dead last among the forty states in which the testing was performed.

82. Robert M. Hazen and James Trefil, as reported by Robert Pool, "Science Literacy: The Enemy Is Us," *Science* 251 (1991): 266–267. See also Robert M. Hazen and James Trefil, *The Sciences: An Integrated Approach* (New York: John Wiley and Sons, 2000).

83. J. Robert Oppenheimer, "Science and the Human Community," in Charles Frankel, ed., *Issues in University Education* (New York: Harper, 1959), p. 58. I first encountered Oppenheimer's remarks in Clark Kerr, *The Uses of the University* (Cambridge: Harvard University Press, 2001), p. 76. Kerr himself wrote at length about the "fractionalization of the intellectual world" and its deleterious effects on the university and society.

84. Trilling, *Mind in the Modern World*, p. 14.

85. For an essay on the sequence of science instruction, see Leon Lederman, "A Science Way of Thinking," *Education Week* 18, no. 40 (June 16, 1999): 1–3.

86. Frank H. Westheimer, "Are Our Universities Rotten at the Core?" *Science* 236 (1987): 1165–1166. The quotation is on p. 1166.

87. Ibid, p. 1165.

88. Donald Kennedy, "College Science: Pass, No Credit," *Science* 293 (2001): 1557.

89. Westheimer, "Are Our Universities Rotten?" p. 1166.

90. Alfred North Whitehead, as quoted in Bruce Alberts, "Scientists as Science Educators," *Issues in Science and Technology* 10, no. 3 (Spring 1994): 29–32. The quotation is on p. 32.

91. Henry Adams, as quoted in Leo Marx, *Does Technology Drive History?* (Cambridge: MIT Press, 1994), p. 27.

92. Freeman Dyson, *Infinite in All Directions* (New York: Harper and Row, 1988), p. 36.

93. H. G. Wells, *The Outline of History, Being a Plain History of Life and Mankind* (London: George Newnes, 1920), vol. 1, p. 11.

Credits

Index

Staphylococci, 126, 127, 128, 131–132
Stehelin, Dominique, 162–163, 164, 165
Stem cell research, 191, 194–200, 250n23
Strachey, Lytton, 112
Streptococcus, 86, 122, 123, 127; Protosil as cure for, 124
Streptomycin, 130, 243n74
Sulfonamides, 123
Sunlight, 170–171, 176, 179
Superconducting Super Collider, 213, 214–215
"Survival of the fittest." *See* Natural selection
Suttner, Bertha von (née Kinsky), 8–9, 14, 24
Sutton, Walter, 152–153
Swanson, Robert, 238n45
Sweden, monarch of, 13, 14
Swedish Academy, 21
Swedish Academy of Science, 22, 28
Syphilis, 82, 84, 85, 105, 116; bacterium, 106; cure for, 123

Taft, Edgar, 45
Task Force on the Health of Research, 208
Tay-Sachs disease, 110, 187
Tchaikovsky, Pyoter Ilich, 103
Teaching, 56, 74; of science, 67–70, 205–207, 226–229
Temin, Howard, 17–18, 52–53, 61
Thalassemia, 109, 190–191
Thomas, Lewis, 223
Tobacco, 146, 148, 179, 210
Tobacco industry, 17–18, 210
Todaro, George, 162
Tolstoy, Leo, 27
Tomkins, Gordon, 35
Toxic shock syndrome, 131
Toxins, 104
Trefil, James, 226

Treponematoses, 106
Trilling, Lionel, 224, 226
Tuberculosis, 82, 94, 104, 110–111; germs, 96; vaccine, 96; drug treatments for, 130; mortality rates for, 131
Tumor(s), 135, 140–141; malignant, 144, 154, 172; extracts of, to transmit cancer, 156–157, 158–159; viruses, 157, 158, 160, 167; sarcomas, 161; lymphatic, 167; mutations in, 168; inherited, 171; retinal, 171; benign, 175; cells, 178
Tumorigenesis, 154–155, 168, 172; genes contributing to, 170, 174–175; mutations and, 171
Tumor suppressor genes, 169–170, 171–172, 173, 247n33; damage to, 176; *TSP53*, 176; mutations in, 178, 247n34
Typhoid fever, 102, 110
Typhus, 82

Ulcers, 131
Ultraviolet radiation, 170–171, 176
United Nations, 23
University of California, 73–74
University of California, San Francisco (UCSF), 72–73, 183–184, 186, 248n3; Bishop at, 51–52; Bishop as Chancellor of, 71–72, 73, 74
University of Pennsylvania, 42, 43, 44

Vaccines, 23, 83, 132, 209–210; for infectious diseases, 91; to control infectious disease, 96; tuberculosis, 96; influenza, 99, 211, 216; effect on infectious disease, 110; smallpox, 115–116, 117, 118, 242n67; microbes and, 115–118; arguments against use of, 118; rabies, 119, 120, 121, 122; attenuated, 119–120; lack of,

21 Prizes not major checkraise
Janice Wagner

49 need for winter
51 B. Shop a different
54 Be a risk taker

01 science as sweet ²37 Rosemary
enterprise

240 foster biography

240 Koch

126 Fleming

243,244 W axman Prize
189 genes have different functions — genes + proteins

Stem Cell Research p194